Besik Chikvinidze

Backward Stochastic Differential Equations and BMO martingales

Besik Chikvinidze

Backward Stochastic Differential Equations and BMO martingales

LAP LAMBERT Academic Publishing

Impressum / Imprint
Bibliografische Information der Deutschen Nationalbibliothek: Die Deutsche Nationalbibliothek verzeichnet diese Publikation in der Deutschen Nationalbibliografie; detaillierte bibliografische Daten sind im Internet über http://dnb.d-nb.de abrufbar.
Alle in diesem Buch genannten Marken und Produktnamen unterliegen warenzeichen-, marken- oder patentrechtlichem Schutz bzw. sind Warenzeichen oder eingetragene Warenzeichen der jeweiligen Inhaber. Die Wiedergabe von Marken, Produktnamen, Gebrauchsnamen, Handelsnamen, Warenbezeichnungen u.s.w. in diesem Werk berechtigt auch ohne besondere Kennzeichnung nicht zu der Annahme, dass solche Namen im Sinne der Warenzeichen- und Markenschutzgesetzgebung als frei zu betrachten wären und daher von jedermann benutzt werden dürften.

Bibliographic information published by the Deutsche Nationalbibliothek: The Deutsche Nationalbibliothek lists this publication in the Deutsche Nationalbibliografie; detailed bibliographic data are available in the Internet at http://dnb.d-nb.de.
Any brand names and product names mentioned in this book are subject to trademark, brand or patent protection and are trademarks or registered trademarks of their respective holders. The use of brand names, product names, common names, trade names, product descriptions etc. even without a particular marking in this works is in no way to be construed to mean that such names may be regarded as unrestricted in respect of trademark and brand protection legislation and could thus be used by anyone.

Coverbild / Cover image: www.ingimage.com

Verlag / Publisher:
LAP LAMBERT Academic Publishing
ist ein Imprint der / is a trademark of
OmniScriptum GmbH & Co. KG
Heinrich-Böcking-Str. 6-8, 66121 Saarbrücken, Deutschland / Germany
Email: info@lap-publishing.com

Herstellung: siehe letzte Seite /
Printed at: see last page
ISBN: 978-3-659-50947-6

Zugl. / Approved by: Tbilisi, Tbilisi State University, Diss., 2013

Copyright © 2014 OmniScriptum GmbH & Co. KG
Alle Rechte vorbehalten. / All rights reserved. Saarbrücken 2014

Tbilisi Ivane Javakhishvili State University

Besik   Chikvinidze

The Faculty of Exact and Natural Sciences

Department of Mathematics

Backward Stochastic Differential Equations and BMO Martingales

Dissertation   Thesis

Scientific Supervisor: Michael Mania

Tbilisi   2013

Table of Contents

1. Introduction 3
2. **New proofs of some classical results on BMO martingales using BSDEs** 10
   2.1 Reverse Hölder and Muckenhoupt conditions and relations with BSDEs 10
   2.2 Girsanov's transformation of BMO martingales and BSDEs 16
3. **Backward stochastic differential equations with a convex generator** 20
   3.1 Value process as a supersolution of equation (1) 20
   3.2 An existence of a solution in the case of bounded characteristics.................. 23
   3.3 The case where the generator f and η are nonnegative 27
   3.4 The general case: the proof of the Theorem 3.1 34
   3.5 The multidimensional case for equation (1) 36
   3.6 Uniqueness of the solution 37
   3.7 Appendix: Some auxiliary assertions 45
4. **Application to the linear-quadratic regulator problem and relation with Bellman-Chitashvili equation** 48
5. **References** 52

1 Introduction

The theory of Backward Stochastic Differential Equations (BSDEs) is a useful tool to study the problem of pricing and hedging of financial derivatives and to give constructions of optimal strategies in stochastic optimal control. These equations have been introduced by Bismut (1978) [4] for the linear case as the equations for the adjoin process in the stochastic maximum principle. The nonlinear BSDE with Bellman type generators was first studied by Chitashvili (1983) [9], who derived a stochastic version of the Bellman equation in optimal control problem. Pardoux and Peng (1990) [32] introduced BSDEs with general generators and proved an existence and uniqueness of the solution under the Lipschitz condition on the driver. Actually, this type of equations appears in several problems of mathematical finance and stochastic optimal control, but in most applications several extensions of the BSDE theory are needed. In various fields of application such as problems of investing and hedging, dynamic financial risk measures, or risk sensitive control problems the Lipschitz continuous framework is not sufficient and an interest for quadratic BSDEs appeared. The question of existence and uniqueness of solutions to these quadratic equations was first examined by Kobylanski (1997) [22] and Lepeltier and San-Martin (1998) [25] for bounded terminal conditions and in Brownian filtration setting. These results were extended for continuous filtrations by Morlais (2008) [31] and Tevzadze (2008) [33]. While Kobylanski's proof can not be generalized to multidimensional case (the proofs by Morlais are similar to [22]), Tevzadze (2008) [33] presents an alternative derivation of Kobylanski's results via a fix point argument. This yields as a byproduct an existence and uniqueness result also for multidimensional case if the generator is quadratic and the norm of the terminal variable is sufficiently small. Recently, Briand and Hu (2009) [5] generalized the existence result for one-dimensional quadratic BSDEs with unbounded boundary condition in case of Brownian filtration. F. Delbaen, Y. Hu and A. Richou (2011) [11] considered BSDEs with convex generators of quadratic growth in Brownian setting. Using the Legandre-Fenchel transform of convex function they showed that any solution of such an equation can be expressed as a value function of a certain optimization problem, which proves the uniqueness of the solution.

In the thesis we studied BSDEs with convex generators of quadratic growth. Existence and uniqueness of a solution is proved for such equations driven by a continuous martingale with unbounded characteristic (**Theorem 3.1** and **Theorem 3.2**). The main novelty of this result is the proof of the existence of a solution for BSDEs driven by continuous martingales with a unbounded characteristics, where we show that the value function of a certain optimization problem satisfies equation (1). Note that our optimization problem and the class of admissible strategies differ from the problem defined in [11], but the resulting value functions coincide. In contrast to [11], we use a linear envelope of convex functions and the optimal strategy of our problem does not exists in general, which enable us to prove the existence result for wider class of generators. This can be viewed as a method of proving that the value process related to a non-Markov optimization problem satisfies the corresponding stochastic Bellman equation, also in the case when an optimal control may not exist. We also proved that under some additional assumptions the value process is the unique solution of equation (1) (**Theorem 3.3**).

We use our results on the existence and uniqueness for BSDEs with quadratic growth to solve the linear-quadratic regulator (LQR) problem in general martingale setting. We derive the corresponding BSDE for LQR problem and express the optimal strategy of LQR problem in terms of the unique solution of corresponding BSDE (**Theorem 4.1** and **Theorem 4.1'**).

In our approach we essentially use properties of Bounded Mean Oscillation (BMO) martingales and estimates of the BMO norms. The BMO martingale theory is extensively used to study backward stochastic differential equations (BSDEs). Some properties of BMO martingales was already used by Bismut [4] when he discussed the existence and uniqueness of a solution of some particular backward stochastic Riccati equations, choosing the BMO space for the martingale part of the solution process. In the work of Delbaen et al [12] conditions for the closedness of stochastic integrals with respect to semimartingales in L^2 were established in relation to the problem of hedging contingent claims and linear BSDEs. Most of this conditions deal with BMO martingales and reverse Hölder inequalities. BMO martingales naturally arise in BSDEs with quadratic generators. When the generator of a BSDE has quadratic growth then the martingale part of any bounded solution of the BSDE is a BMO martingale. This fact was proved in [19, 22, 26, 27, 29, 33] under various degrees of generality. Later, the BMO norms were used to prove an existence, uniqueness and stability results for BSDEs, among others in [1, 2, 6, 7, 18, 28, 31, 33].

As it was mentioned above when the generator of a BSDE has quadratic growth then the martingale part of any bounded solution of the BSDE is a BMO martingale, which implies that the stochastic exponent of the martingale part is a uniformly integrable martingale. Note that for such BSDEs, martingale parts of unbounded solutions are not BMO martingales in general. But we have proved that if a solution has an exponential moments, then the stochastic exponent of martingale part of the solution is a uniformly integrable martingale (**Proposition 3.7**). Note that similar result was obtained in Mocha and Westrey [30]

In **Chapter 2** we provide new proofs of some classical results on BMO martingales: using the BSDE tool, we first provide a new characterization of both the Reverse Holder inequality and of the Muckenhoupt inequality which are equivalent to the BMO martingale property. Those last properties implying the uniform integrability of the associate stochastic exponential, the Girsanov's transform induces a true change of measure. Using these BSDE characterizations, we recover the fact that the Girsanov's transform is an isomorphism between two BMO spaces if the BMO property of the stochastic exponential holds. Another achievement is that we also improve the constants in the inequalities comparing the two BMO norms associated with the previous isomorphism.

To formulate main results of the thesis we shall introduce basic objects and give some definitions.

Let a basic probability space with filtration $(\Omega, \mathcal{F}, \{\mathcal{F}_t\}_{0 \le t \le T}, P)$ be given. Suppose that the flow of σ-algebras $\{\mathcal{F}_t\}_{0 \le t \le T}$ is complete and right continuous.

We consider a backward stochastic differential equation ($BSDE$) of the form

$$\begin{cases} Y_t = Y_0 - \int_0^t f(s, Z_s) d\langle M \rangle_s + \int_0^t Z_s dM_s + L_t, \\ Y_T = \eta, \end{cases} \quad (1)$$

where the generator $f : [0;T] \times \Omega \times R \longrightarrow R$ is a measurable function and for any z, $f(\cdot, \cdot, z)$ is a predictable process; η is an $\mathcal{F}_T$-measurable random variable and $\{M_t\}_{0 \le t \le T}$ is a given square integrable martingale with respect to the filtration $\{\mathcal{F}_t\}_{0 \le t \le T}$. The pair (f, η) is called the parameters of equation (1).

Definition 1.1. A solution of equation (1) is a triple (Y, Z, L), where $\{Y_t\}_{0 \le t \le T}$ is a semimartingale, $\{Z_t\}_{0 \le t \le T}$ is a predictable M integrable process, $\{L_t\}_{0 \le t \le T}$ is a local martingale orthogonal to M and the triple (Y, Z, L) satisfies equation (1).

Sometimes we shall say that the solution is only the first component of the triple (Y, Z, L) keeping in mind that $\int Z dM + L$ is the martingale part of Y.

Now let us recall definitions of BMO martingales, Reverse Hölder and Muckenhoupt conditions:

Definition 1.2. We say that a continuous martingale M is from the class BMO if

$$\sup_\tau \left\| E[\langle M \rangle_T - \langle M \rangle_\tau | \mathcal{F}_\tau] \right\|_\infty < \infty,$$

where the supremum is taken over all stopping times τ, $0 \leq \tau \leq T$.

The process $\mathcal{E}_t(M) = \exp(M_t - \frac{1}{2}\langle M \rangle_t)$ is called the stochastic exponent from the martingale M. Denote $\mathcal{E}_{t,T}(M) := \mathcal{E}_T(M)/\mathcal{E}_t(M)$.

It is well known that if M is a local martingale then $\mathcal{E}(M)$ is a local martingale too. Kazamaki [21] proved that if a martingale M is from the class BMO then $\mathcal{E}(M)$ is a uniformly integrable martingale. Therefore $\mathcal{E}(M)$ defines a new probability measure defined by

$$d\tilde{P} = \mathcal{E}_T(M)dP.$$

Let $\tilde{M} = \langle M \rangle - M$ which according to the Girsanov theorem is a local $\tilde{P}$ martingale.

Definition 1.3. Let $1 < p < \infty$. $\mathcal{E}(M)$ is said to satisfy the reverse Hölder condition (R_p), if the inequality

$$E\left[\{\mathcal{E}_{\tau,T}(M)\}^p \big| \mathcal{F}_\tau \right] \leq C_p$$

is valid for every stopping time τ, with a constant $C_p > 0$ depending only on p.

If $\mathcal{E}(M)$ is a uniformly integrable martingale then by the Jensen inequality we also have that $E\left[\{\mathcal{E}_{\tau,T}(M)\}^p \big| \mathcal{F}_\tau \right] \geq 1$.

A condition dual to (R_p) is the Muckenhoupt condition (A_p).

Definition 1.4. $\mathcal{E}(M)$ is said to satisfy (A_p) condition for $1 < p < \infty$ if there is a constant $D_p > 0$ such that for every stopping time $\tau \in [0, T]$

$$E\left[\{\mathcal{E}_{\tau,T}(M)\}^{-\frac{1}{p-1}} \big| \mathcal{F}_\tau \right] \leq D_p.$$

Note that, since $\mathcal{E}(M)$ is a supermartingale, the Jensen inequality implies the converse inequality

$$E\left[\{\mathcal{E}_{\tau,T}(M)\}^{-\frac{1}{p-1}} \big| \mathcal{F}_\tau \right] \geq \left\{ E[\mathcal{E}_{\tau,T}(M) | \mathcal{F}_\tau] \right\}^{-\frac{1}{p-1}} \geq 1.$$

The following Lemma from **Chapter 2** establishes a new characterization of the reverse Hölder and Muckenhoupt conditions in terms of BSDE.

Lemma 2.1 Let M be a continuous local martingale.
a) $\mathcal{E}(M)$ satisfies (R_p) if and only if there exists a bounded, positive solution of BSDE

$$\begin{cases} Y_t = Y_0 - \int_0^t [\frac{p(p-1)}{2} Y_s + p\psi_s] d\langle M \rangle_s + \int_0^t \psi_s dM_s + N_t, \\ Y_T = 1. \end{cases} \quad (2)$$

b) $\mathcal{E}(M)$ satisfies (A_p) if and only if there exists a bounded, positive solution of equation

$$\begin{cases} X_t = X_0 - \int_0^t [\frac{p}{2(p-1)^2} X_s - \frac{1}{p-1}\varphi_s] d\langle M \rangle_s + \int_0^t \varphi_s dM_s + L_t, \\ X_T = 1. \end{cases} \quad (3)$$

Using properties of BSDEs we prove the well known equivalence between BMO property, Muckenhoupt and reverse Hölder conditions (Doleans-Dade and Meyer [15], Kazamaki [21]) and obtain BMO norm estimates in terms of reverse Hölder and Muckenhaupt constants. These results is given in **Theorem 2.1** from **Chapter 2**:

Theorem 2.1 Let $\mathcal{E}(M)$ be a uniformly integrable martingale. Then the following conditions are equivalent:
i). $\tilde{M} \in BMO(\tilde{P})$.
ii). $\mathcal{E}(M)$ satisfies the (R_p) condition for some $p > 1$.
iii). $M \in BMO(P)$.
iv). $\mathcal{E}(M)$ satisfies the (A_p) condition for some $p > 1$.

It is well known that if M is a BMO martingale, then the mapping $\phi : \mathcal{L}(P) \ni X \longrightarrow \tilde{X} = \langle X, M \rangle - X \in \mathcal{L}(\tilde{P})$ is an isomorphism of $BMO(P)$ onto $BMO(\tilde{P})$, where $d\tilde{P} = \mathcal{E}_T(M)dP$. E. g., it was proved by Kazamaki [20, 21] that the inequality

$$\|\tilde{X}\|_{BMO(\tilde{P})} \leq C_K(\tilde{M}) \cdot \|X\|_{BMO(P)} \qquad (4)$$

is valid for all $X \in BMO(P)$, where the constant $C_K(\tilde{M}) > 0$ is independent of X but depends on the martingale M. Using the properties of a suitable BSDE we prove this inequality with a constant $C(\tilde{M})$ which we express as a linear function of the $BMO(\tilde{P})$ norm of $\tilde{M} = \langle M \rangle - M$ and which is less than $C_K(\tilde{M})$ for all values of this norm. This result is given in **Theorem 2.2** of **Chapter 2**:

Theorem 2.2 If $M \in BMO(P)$, then $\phi : X \to \tilde{X}$ is an isomorphism of $BMO(P)$ onto $BMO(\tilde{P})$. In particular, the inequality

$$\frac{1}{\left(1 + \frac{\sqrt{2}}{2}\|M\|_{BMO(P)}\right)}\|X\|_{BMO(P)} \leq \|\tilde{X}\|_{BMO(\tilde{P})} \leq \left(1 + \frac{\sqrt{2}}{2}\|\tilde{M}\|_{BMO(\tilde{P})}\right)\|X\|_{BMO(P)} \qquad (5)$$

is valid for any $X \in BMO(P)$.

Let us compare the constant

$$C(\tilde{M}) = 1 + \frac{\sqrt{2}}{2}\|\tilde{M}\|_{BMO(\tilde{P})}$$

with the corresponding constant $C_K(\tilde{M})$ from (4) (Kazamaki [20, 21]). Since

$$E^{\tilde{P}}\left[\{\mathcal{E}_{\tau,T}(\tilde{M})\}^{-\frac{1}{p-1}}\Big|\mathcal{F}_\tau\right] \geq 1,$$

the constant $C_K(\tilde{M})$ is more than $\sqrt{2p}$, where p is such that $\|\tilde{M}\|_{BMO(\tilde{P})} < \sqrt{2}(\sqrt{p}-1)$. Since the last inequality is equivalent to the inequality $p > \left(1 + \frac{\sqrt{2}}{2}\|\tilde{M}\|_{BMO(\tilde{P})}\right)^2$, we obtain that at least

$$C^2(\tilde{M}) \leq \frac{1}{2}C_K^2(\tilde{M}).$$

It is evident that in the trivial case $M = 0$, we have that $\tilde{P} = P$ and $\tilde{X} = X$. Note that, if $M = 0$ then (5) gives the two-sided inequality

$$||X||_{BMO(P)} \leq ||\tilde{X}||_{BMO(\tilde{P})} \leq ||X||_{BMO(P)},$$

implying the equality $\tilde{X} = X$, whereas from (4), we only have

$$\frac{1}{2}||X||_{BMO(P)} \leq ||\tilde{X}||_{BMO(\tilde{P})} \leq 2||X||_{BMO(P)}.$$

This shows that the following simple corollary cannot be deduced from inequality (4).

Corollary. Let $(M^n, n \geq 1)$ be a sequence of $BMO(P)$ martingales such that $\lim_{n \to \infty} ||M^n||_{BMO(P)} = 0$. Let $dP^n = \mathcal{E}_T(M^n)dP$ and $\tilde{X}^n = \langle X, M^n \rangle - X$. Then for any $X \in BMO(P)$

$$\lim_{n \to \infty} ||\tilde{X}^n||_{BMO(P^n)} = ||X||_{BMO(P)}.$$

In **Chapter 3** of the thesis we consider the backward stochastic differential equations (BSDEs) of type (1) with convex generators of quadratic growth and prove the existence and uniqueness of the solution when the martingale characteristic and terminal value admit exponential moments of some order. This result is given in **Theorem 3.1**:

Theorem 3.1 Suppose that the filtration $\{\mathcal{F}_t\}_{0 \leq t \leq T}$ is continuous and $\{M_t\}_{0 \leq t \leq T}$ is a martingale from the class BMO. Let the parameters (f, η) of equation (1) satisfy the following conditions:
1) for any (t, ω), $f(t, \omega, \cdot)$ is a continuous and convex function.
2) there exist a predictable non-negative process α_t and a constant $\gamma \geq 0$ such that $\int \alpha dM \in BMO$ and for any (t, ω, z)

$$|f(t, \omega, z)| \leq \alpha_t(\omega) + \frac{\gamma}{2} z^2.$$

3) $E e^{\gamma \eta + \gamma \int_0^T \alpha_s d\langle M \rangle_s} < \infty$ and $\eta + \int_0^T f(s, 0) d\langle M \rangle_s \geq -C$ for some $C \geq 0$.
Then there exists a solution $V = \{V_t\}_{0 \leq t \leq T}$ of equation (1) represented in the form

$$V_t = \operatorname*{ess\,sup}_{u \in U} E\left[\mathcal{E}_{t,T}\left(\int f_l'(u)dM\right)\left(\eta + \int_t^T [f(s, u_s) - f_l'(s, u_s)u_s]d\langle M \rangle_s\right)\Big|\mathcal{F}_t\right]$$

where U is the class of predictable bounded controls.

The main idea of the proof is as follows: since f is convex, according to Lemma 3.16 (see Appendix) we have the equality

$$f(t, Z_t) = \operatorname*{ess\,sup}_{u \in U}[f(t, u_t) + f_l'(t, u_t)(Z_t - u_t)].$$

Let us consider the linear (BSDE)

$$\begin{cases} Y_t = Y_0 - \int_0^t [f(s, u_s) + f_l'(s, u_s)(Z_s - u_s)]d\langle M \rangle_s + \int_0^t Z_s dM_s + L_t, \\ Y_T = \eta, \end{cases} \quad (6)$$

For any $u \in U$ equation (6) admits a unique solution (Y^u, Z^u, L^u) where the first component Y^u has the form
$$Y_t^u = E^u\left[\eta + \int_t^T [f(s, u_s) - f_l'(s, u_s)u_s]d\langle M\rangle_s \Big| \mathcal{F}_t\right].$$
Here E^u denotes the conditional expectation with respect to the measure $dP^u = \mathcal{E}_T\left(\int f_l'(u)dM\right)dP$. Our task is to prove that the value process
$$V_t = \operatorname{ess\,sup}_{u \in U} Y_t^u$$
is the solution of equation (1).

We shall show this first in the case when η, $\langle M\rangle_T$ and $\int_0^T \alpha_s^2 d\langle M\rangle_s$ are bounded random variables and then we complete the proof of Theorem 3.1 by passing to the limit.

As mentioned above, the uniqueness of a solution of BSDEs with convex generators of quadratic growth was proved by F. Delbaen, Y. Hu and A. Richou (2009) [11] for Brownian filtration. Using similar method we prove the uniqueness result for continuous martingales with unbounded characteristics.

The uniqueness of the solution of equation (1) we prove in the class of processes:
$$\aleph = \left\{(Y, Z, L) : Ee^{p(Y^+ + \int \alpha d\langle M\rangle)^*} < \infty \text{ and } Ee^{\epsilon(Y^-)^*} < \infty\right\}$$
for some $p > \gamma$ and $\epsilon > 0$.

Theorem 3.3 If in addition to the conditions of **Theorem 3.1**
$$Ee^{p(\eta^+ + \int \alpha d\langle M\rangle)^*} < \infty$$
where $p > \gamma$, then the value process V is the unique solution of equation (1) in the class $\aleph$.

The existence and uniqueness theorems (Th 3.1 and Th 3.2) we shall use to solve the linear-quadratic regulator problem (LQR) in general martingale setting (see **Chapter 4**). More precisely: Let A be a decision set and let M be continuous local martingale. To any $a \in A$ we associate a local martingale $M^a = aM$. Controls are predictable mappings $u : \Omega \times [0; T] \longrightarrow A$ and probability measures P^u corresponding to any control u are defined by $dP^u = \mathcal{E}_T(M^u)dP$, where $M_t^u = \int_0^t u_s dM_s$, provided that $\mathcal{E}_t(M^u)$ is a uniformly integrable martingale. Assume that the cost criterium is of the form $r(t, a) = -g(t)a^2 + h(t)$ and consider an optimization problem to maximize
$$E^u\left[\eta + \int_0^T (h(s) - g(s)u_s^2)d\langle M\rangle_s\right]$$
over all $u \in U$, where U is a class of admissible strategies and η is a $\mathcal{F}_T$-measurable random variable. Let
$$V_t = \operatorname{ess\,sup}_{u \in U} E^u\left[\eta + \int_t^T (h(s) - g(s)u_s^2)d\langle M\rangle_s \Big| \mathcal{F}_t\right]$$
be the value process of the problem. We derive the corresponding BSDE for the value process:

$$Y_t = Y_0 - \int_0^t [h(s) + \frac{1}{4g(s)} Z_s^2] d\langle M \rangle_s + \int_0^t Z_s dM_s + L_t, Y_T = \eta. \quad (7)$$

Note that in this case the condition: $(\operatorname{ess\,sup}_{u \in U} \langle M^u \rangle)_t$ is locally bounded is not satisfied and the existence of a solution of corresponding BSDE does not follows from Chitashvili's [9] result. But we can use the above mentioned theorems of existence and uniqueness of the solution. Under the conditions:

(i) M, $\int |h| dM \in BMO$

(ii) $Ee^{p(\eta^+ + \int_0^T |h(s)| d\langle M \rangle_s)^*} < \infty$ and $\eta + \int_0^T |h(s)| d\langle M \rangle_s \geq -D > -\infty$

where $p > \frac{1}{2\varepsilon}$ and $g(s) \geq \varepsilon > 0$ for some $\varepsilon > 0$, it follows from **Theorem 3.1** and **Theorem 3.3** that there exists the unique solution of equation (7) $(\tilde{V}, Z, L)$, where the first component $\tilde{V}$ coincides with the value process V and the optimal strategy is equal to $\frac{Z}{2g}$ (see **Theorem 4.1**).

Remark: Notice that for simplicity we have considered the case when $M^a = aM$. The similar result can be obtained if for any $a \in A$ we associate a local martingale $M^a = M^0 + aM$, where M^0 is a fixed local martingale.

2 New proofs of some classical results on BMO martingales using BSDEs

We start with a probability space $(\Omega, \mathcal{F}, P)$, a finite time horizon $0 < T < \infty$ and a filtration $F = (\mathcal{F}_t)_{0 \leq t \leq T}$ satisfying the usual conditions of right-continuity and completeness.

Throughout this chapter we shall assume that M is a continuous local martingale with $\langle M \rangle_T < \infty$ P- a.s. This implies that $\mathcal{E}_t(M) > 0$ P-a.s. for all $t \in [0, T]$, which allows to define $\mathcal{E}_{\tau,T}(M)$ as $\mathcal{E}_{\tau,T}(M) = \mathcal{E}_T(M)/\mathcal{E}_\tau(M)$.

2.1 Reverse Hölder and Muckenhoupt conditions and relations with BSDEs

We recall definitions of BMO martingales, Reverse Hölder and Muckenhaupt conditions (see, e.g., Doleans-Dade and Meyer [15], or Kazamaki [21]).

Definition 2.1. Let $1 < p < \infty$. $\mathcal{E}(M)$ is said to satisfy (R_p) condition if the reverse Hölder inequality

$$E\Big[\{\mathcal{E}_{\tau,T}(M)\}^p \Big| \mathcal{F}_\tau\Big] \leq C_p$$

is valid for every stopping time τ, with a constant $C_p > 0$ depending only on p.

If $\mathcal{E}(M)$ is a uniformly integrable martingale then by the Jensen inequality we also have that $E\Big[\{\mathcal{E}_{\tau,T}(M)\}^p \Big| \mathcal{F}_\tau\Big] \geq 1$.

A condition dual to (R_p) is the Muckenhoupt condition (A_p).

Definition 2.2. $\mathcal{E}(M)$ is said to satisfy (A_p) condition for $1 < p < \infty$ if there is a constant $D_p > 0$ such that for every stopping time $\tau \in [0, T]$

$$E\Big[\{\mathcal{E}_{\tau,T}(M)\}^{-\frac{1}{p-1}} \Big| \mathcal{F}_\tau\Big] \leq D_p.$$

Note that, since $\mathcal{E}(M)$ is a supermartingale, the Jensen inequality implies the converse inequality

$$E\Big[\{\mathcal{E}_{\tau,T}(M)\}^{-\frac{1}{p-1}} \Big| \mathcal{F}_\tau\Big] \geq \Big\{E[\mathcal{E}_{\tau,T}(M)|\mathcal{F}_\tau]\Big\}^{-\frac{1}{p-1}} \geq 1.$$

In this section we shall consider only linear BSDEs of the type

$$\begin{cases} Y_t = Y_0 - \int_0^t [\alpha Y_s + \beta \psi_s] d\langle M \rangle_s + \int_0^t \psi_s dM_s + N_t, \\ Y_T = 1, \end{cases} \quad (8)$$

where α and β are constants.

Now we give a new characterization of reverse Hölder and Muckenhoupt conditions in terms of corresponding BSDEs.

Lemma 2.1. *Let M be a continuous local martingale.*
a) $\mathcal{E}(M)$ *satisfies* (R_p) *if and only if there exists a bounded, positive solution of BSDE*

$$\begin{cases} Y_t = Y_0 - \int_0^t [\frac{p(p-1)}{2} Y_s + p\psi_s] d\langle M \rangle_s + \int_0^t \psi_s dM_s + N_t, \\ Y_T = 1. \end{cases} \quad (9)$$

b) $\mathcal{E}(M)$ *satisfies* (A_p) *if and only if there exists a bounded, positive solution of equation*

$$\begin{cases} X_t = X_0 - \int_0^t [\frac{p}{2(p-1)^2} X_s - \frac{1}{p-1} \varphi_s] d\langle M \rangle_s + \int_0^t \varphi_s dM_s + L_t, \\ X_T = 1. \end{cases} \quad (10)$$

Proof: **a)** Let first show that if $\mathcal{E}(M)$ satisfies (R_p) then the process $Y_t = E\left[\{\mathcal{E}_{t,T}(M)\}^p \big| \mathcal{F}_t\right]$ is a solution of BSDE (9). It is evident that Y is a bounded positive process and that $Y_t\{\mathcal{E}_t(M)\}^p$ is a uniformly integrable martingale. Therefore, since $\mathcal{E}_t(M) > 0$, the process Y will be a special semimartingale. Let $Y_t = Y_0 + A_t + m_t$ be the canonical decomposition of Y, where m is a locally square integrable martingale and A a predictable process of bounded variation. Using the Galtchouk-Kunita-Watanabe decomposition for m, we get

$$Y_t = Y_0 + A_t + \int_0^t \psi_s dM_s + N_t, \quad (11)$$

where N is a local martingale orthogonal to M.

Now using the Itô formula we have

$$Y_t \{\mathcal{E}_t(M)\}^p = Y_0 + \int_0^t [\frac{p(p-1)}{2} Y_s + p\psi_s] \{\mathcal{E}_s(M)\}^p d\langle M \rangle_s +$$

$$+ \int_0^t \{\mathcal{E}_s(M)\}^p dA_s + \tilde{m}_t, \quad (12)$$

where $\tilde{m}$ is a local martingale.

Because $Y_t\{\mathcal{E}_t(M)\}^p$ is a martingale, equalizing the part of bounded variation to zero, we obtain that

$$A_t = -\int_0^t [\frac{p(p-1)}{2} Y_s + p\psi_s] d\langle M \rangle_s,$$

which implies that $Y_t = E\left[\{\mathcal{E}_{t,T}(M)\}^p \big| \mathcal{F}_t\right]$ is a solution of equation (9).

Now let equation (9) admits a bounded positive solution Y_t. Using the Itô formula for the process $Y_t\{\mathcal{E}_t(M)\}^p$ we get that $Y_t\{\mathcal{E}_t(M)\}^p$ is a local martingale. Hence it is a supermartingale, as a positive local martingale. Therefore, from the supermartingale inequality and the boundary condition $Y_T = 1$ we obtain that $E\left[\{\mathcal{E}_{t,T}(M)\}^p \big| \mathcal{F}_t\right] \leq Y_t$. Because Y is bounded, this implies that $\mathcal{E}(M)$ satisfies (R_p) condition.
b) The proof is similar to the proof of the part a), we only need to replace p by $-\frac{1}{p-1}$. □

Let $\mathcal{E}(M)$ be a uniformly integrable martingale. Denote by $\tilde{P}$ a new probability measure defined by $d\tilde{P} = \mathcal{E}_T(M) dP$ and let $\tilde{M} = \langle M \rangle - M$.

Now we shall give a new proof of the well known equivalence (Doleans-Dade and Meyer [15], Kazamaki [21]) between BMO property, Muckenhoupt and reverse Hölder conditions.

Theorem 2.1. Let $\mathcal{E}(M)$ be a uniformly integrable martingale. Then the following conditions are equivalent:
i). $\tilde{M} \in BMO(\tilde{P})$.
ii). $\mathcal{E}(M)$ satisfies the (R_p) condition for some $p > 1$.
iii). $M \in BMO(P)$.
iv). $\mathcal{E}(M)$ satisfies the (A_p) condition for some $p > 1$.

Proof: For the sake of simplicity, in all proofs given here, we shall assume without loss of generality that all stochastic integrals are martingales, otherwise one can use the localization arguments. Before we prove the theorem define the special space of solutions of (8): A solution of such a BSDE is defined as a triple (Y, ψ, N), with $\langle N, M \rangle = 0$, from the space $S^\infty \times BMO(P) \times H^2(P)$ equipped with the following norms

$$\|Y\|_\infty = \|Y_T^*\|_{L^\infty}, \quad \text{where} \quad Y_T^* = \sup_{t \in [0,T]} |Y_t|,$$

$$\|\psi \cdot M\|_{BMO(P)} = \sup_\tau \left\| E\left[\int_\tau^T \psi_s^2 d\langle M \rangle_s \Big| \mathcal{F}_\tau \right]^{1/2} \right\|_\infty,$$

$$\|N\|_{H^2} = E^{\frac{1}{2}}[N]_T,$$

where $[N]$ is the square bracket of N.

Note that, since the martingale M is assumed to be continuous, only the latter term of the equation (8) may have the jumps, i.e., $\Delta Y = \Delta N$. In order to avoid the definition of BMO norms for right-continuous martingales, we are using the H^2 norms for orthogonal martingale parts. This is sufficient for our goals, since the generators of equations under consideration does not depend on orthogonal martingale parts.

$i) \Longrightarrow ii)$

Let $\tilde{M} \in BMO(\tilde{P})$. According to Lemma 2.1 it is sufficient to show that equation (9) admits a bounded positive solution for some $p > 1$. Let us rewrite equation (9) in terms of the $\tilde{P}$-martingale $\tilde{M}$:

$$\begin{cases} Y_t = Y_0 - \int_0^t \left[\frac{p(p-1)}{2} Y_s + (p-1)\psi_s \right] d\langle \tilde{M} \rangle_s - \int_0^t \psi_s d\tilde{M}_s + N_t, \\ Y_T = 1, \end{cases} \quad (13)$$

Since $\langle N, M \rangle = 0$, N is a local $\tilde{P}$- martingale orthogonal to $\tilde{M}$.

Define the mapping $H : S^\infty \times BMO(\tilde{P}) \times H^2(\tilde{P})$ into itself, which maps $(y, \psi, n) \in S^\infty \times BMO(\tilde{P}) \times H^2(\tilde{P})$ onto the solution (Y, Ψ, N) of the BSDE (13), i.e.,

$$Y_t = E^{\tilde{P}}\left[1 + \int_t^T \left[\frac{p(p-1)}{2} y_s + (p-1)\psi_s \right] d\langle \tilde{M} \rangle_s \Big| \mathcal{F}_t \right]$$

and

$$-\int_0^t \Psi_s d\tilde{M}_s + N_t = E^{\tilde{P}}\left[1 + \int_0^T \left[\frac{p(p-1)}{2} y_s + (p-1)\psi_s \right] d\langle \tilde{M} \rangle_s \Big| \mathcal{F}_t \right] -$$
$$- E^{\tilde{P}}\left[1 + \int_0^T \left[\frac{p(p-1)}{2} y_s + (p-1)\psi_s \right] d\langle \tilde{M} \rangle_s \right].$$

We shall show that there exists $p > 1$ such that this mapping is a contraction.

Let
$$\delta Y = Y^1 - Y^2,\ \delta y = y^1 - y^2,\ \delta\Psi = \Psi^1 - \Psi^2,\ \delta\psi = \psi^1 - \psi^2,\ \delta N = N^1 - N^2.$$
It is evident that $\delta Y_T = 0$ and
$$\delta Y_t = \delta Y_0 - \int_0^t \Big[\frac{p(p-1)}{2}\delta y_s + (p-1)\delta\psi_s\Big]d\langle \tilde M\rangle_s - \int_0^t \delta\Psi_s d\tilde M_s + \delta N_t.$$
Applying the Itô formula to $(\delta Y_\tau)^2 - (\delta Y_T)^2$ and taking conditional expectations we have
$$(\delta Y_\tau)^2 + E^{\tilde P}\Big[\int_\tau^T (\delta\Psi_s)^2 d\langle\tilde M\rangle_s\Big|\mathcal F_\tau\Big] + E^{\tilde P}\Big[[\delta N]_T - [\delta N]_\tau\Big|\mathcal F_\tau\Big] =$$
$$= E^{\tilde P}\Big[\int_\tau^T p(p-1)\delta Y_s \delta y_s d\langle\tilde M\rangle_s\Big|\mathcal F_\tau\Big] + E^{\tilde P}\Big[\int_\tau^T 2(p-1)\delta Y_s\delta\psi_s d\langle\tilde M\rangle_s\Big|\mathcal F_\tau\Big]$$
and using elementary inequalities we obtain
$$(\delta Y_\tau)^2 + E^{\tilde P}\Big[\int_\tau^T(\delta\Psi_s)^2 d\langle\tilde M\rangle_s\Big|\mathcal F_\tau\Big] + E^{\tilde P}\Big[[\delta N]_T - [\delta N]_\tau\Big|\mathcal F_\tau\Big] \le$$
$$\le \frac{p(p-1)}{2}\|\tilde M\|^2_{BMO(\tilde P)}\cdot\|\delta Y\|^2_\infty + \frac{p(p-1)}{2}\|\tilde M\|^2_{BMO(\tilde P)}\cdot\|\delta y\|^2_\infty +$$
$$+(p-1)\|\tilde M\|^2_{BMO(\tilde P)}\cdot\|\delta Y\|^2_\infty + (p-1)\Big\|\int\delta\psi d\tilde M\Big\|^2_{BMO(\tilde P)}.$$
Because the right hand side of the inequality does not depend on τ, we will have
$$\|\delta Y\|^2_\infty + \Big\|\int\delta\Psi d\tilde M\Big\|^2_{BMO(\tilde P)} + \|\delta N\|^2_{H^2(\tilde P)} \le$$
$$\le \frac{3p(p-1)}{2}\|\tilde M\|^2_{BMO(\tilde P)}\cdot\|\delta Y\|^2_\infty + \frac{3p(p-1)}{2}\|\tilde M\|^2_{BMO(\tilde P)}\cdot\|\delta y\|^2_\infty +$$
$$+3(p-1)\|\tilde M\|^2_{BMO(\tilde P)}\cdot\|\delta Y\|^2_\infty + 3(p-1)\Big\|\int\delta\psi d\tilde M\Big\|^2_{BMO(\tilde P)}.$$
From this we easily obtain the inequality:
$$\Big(1 - \frac{3p(p-1)}{2}\|\tilde M\|^2_{BMO(\tilde P)} - 3(p-1)\|\tilde M\|^2_{BMO(\tilde P)}\Big)\|\delta Y\|^2_\infty +$$
$$+\Big\|\int\delta\Psi d\tilde M\Big\|^2_{BMO(\tilde P)} + \|\delta N\|^2_{H^2(\tilde P)} \le$$
$$\le \frac{3p(p-1)}{2}\|\tilde M\|^2_{BMO(\tilde P)}\|\delta y\|^2_\infty + 3(p-1)\Big\|\int\delta\psi d\tilde M\Big\|^2_{BMO(\tilde P)}. \qquad (14)$$
Since
$$1 - \frac{3}{2}(p-1)(p+2)\|\tilde M\|^2_{BMO(\tilde P)} > 0$$
for p sufficiently close to 1, one can make the constant of $\|\delta Y\|^2_\infty$ in the left-hand side of (14) positive and we finally obtain the inequality
$$\|\delta Y\|^2_\infty + \Big\|\int\delta\Psi d\tilde M\Big\|^2_{BMO(\tilde P)} + \|\delta N\|^2_{H^2(\tilde P)} \le$$

$$\leq \alpha(p) \cdot ||\delta y||_\infty^2 + \beta(p) \cdot \left\| \int \delta \psi d\tilde{M} \right\|_{BMO(\tilde{P})}^2, \tag{15}$$

where

$$\alpha(p) = \frac{3p(p-1)||\tilde{M}||_{BMO(\tilde{P})}^2}{2 - 3(p-1)(p+2)||\tilde{M}||_{BMO(\tilde{P})}^2},$$

$$\beta(p) = \frac{6(p-1)}{2 - 3(p-1)(p+2)||\tilde{M}||_{BMO(\tilde{P})}^2}.$$

It is easy to see that $\lim_{p\downarrow 1} \alpha(p) = \lim_{p\downarrow 1} \beta(p) = 0$. So, if we take p^* such that $\alpha(p^*) < 1$ and $\beta(p^*) < 1$ we obtain that there exists $0 < C < 1$ such that

$$||\delta Y||_\infty^2 + \left\| \int \delta \Psi d\tilde{M} \right\|_{BMO(\tilde{P})}^2 + ||\delta N||_{H^2(\tilde{P})}^2 \leq$$

$$\leq C\left(||\delta y||_\infty^2 + \left\| \int \delta \psi d\tilde{M} \right\|_{BMO(\tilde{P})}^2 + ||\delta n||_{H^2(\tilde{P})}^2\right), \tag{16}$$

for any $(y, \psi, n) \in S^\infty \times BMO(\tilde{P}) \times H^2(\tilde{P})$.

Thus, the mapping H is a contraction and there exists a fixed-point of H (Y, Ψ, N), which is the unique solution of (9) in $S^\infty \times BMO(\tilde{P}) \times H^2(\tilde{P})$.

Since $\alpha(p)$ and $\beta(p)$ are decreasing functions of $p \in (1, \infty)$, the norms $||Y||_\infty$ and $||\Psi \cdot \tilde{M}||_{BMO(\tilde{P})}$ are uniformly bounded, as functions of p for $p \in [1, p^*]$. Therefore, for any $p \in [1, p^*]$ we have

$$Y_t = E^{\tilde{P}}\left[1 + \int_t^T \left[\frac{p(p-1)}{2} Y_s + (p-1)\Psi_s\right] d\langle \tilde{M}\rangle_s \Big| \mathcal{F}_t\right] \tag{17}$$

and

$$Y_t \geq 1 - \frac{p(p-1)}{2}||Y||_\infty ||\tilde{M}||_{BMO(\tilde{P})}^2 - \frac{p-1}{2}||\tilde{M}||_{BMO(\tilde{P})}^2 -$$

$$-\frac{p-1}{2}||\Psi \cdot \tilde{M}||_{BMO(\tilde{P})}^2 \geq 0$$

for some p sufficiently close to 1. Hence, there exists a bounded positive solution of equation (9) for some $p > 1$, which implies that $\mathcal{E}(M)$ satisfies the R_p condition, according to Lemma 2.1.

$ii) \Longrightarrow iii)$ Let $\mathcal{E}(M)$ be a uniformly integrable martingale and satisfies the (R_p) condition for some $p > 1$. Then the process $Y_t = E\left[\{\mathcal{E}_{t,T}(M)\}^p\Big|\mathcal{F}_t\right]$ is a solution of equation (9) and satisfies the two-sided inequality

$$1 \leq Y_t \leq C_p. \tag{18}$$

Applying the Itô formula for $e^{-\beta Y_T} - e^{-\beta Y_\tau}$ and taking conditional expectations we have

$$e^{-\beta} - e^{-\beta Y_\tau} = \beta \frac{p(p-1)}{2} E\left[\int_\tau^T Y_s e^{-\beta Y_s} d\langle M\rangle_s \Big|\mathcal{F}_\tau\right] +$$

$$+ E\left[\int_\tau^T e^{-\beta Y_s}\left(\frac{\beta^2}{2}\psi_s^2 + \beta p \psi_s\right) d\langle M\rangle_s \Big|\mathcal{F}_\tau\right] + \frac{\beta^2}{2} E\left[\int_\tau^T e^{-\beta Y_s} d\langle N^c\rangle_s \Big|\mathcal{F}_\tau\right] +$$

$$+ E\left[\Sigma_{\tau < s \leq T}\left(e^{-\beta Y_s} - e^{-\beta Y_{s-}} + \beta e^{-\beta Y_{s-}} \Delta Y_s\right)\Big|\mathcal{F}_\tau\right].$$

Since $\frac{\beta^2}{2}\psi_s^2 + \beta p\psi_s \geq -\frac{p^2}{2}$ and $e^{-\beta Y_s-} - e^{-\beta Y_s-} + \beta e^{-\beta Y_s-}\Delta Y_s \geq 0$ we obtain the inequality

$$\frac{p}{2}E\left[\int_\tau^T (\beta(p-1)Y_s - p)e^{-\beta Y_s}d\langle M\rangle_s \Big| \mathcal{F}_\tau\right] \leq e^{-\beta} - e^{-\beta Y_\tau}.$$

Then from the two-sided inequality (18) it follows that for any $\beta > \frac{p}{p-1}$

$$\frac{p}{2}(\beta(p-1) - p)e^{-\beta C_p}E\left[\langle M\rangle_T - \langle M\rangle_\tau \Big| \mathcal{F}_\tau\right] \leq e^{-\beta} - e^{-\beta C_p}, \qquad (19)$$

which implies that

$$\|M\|_{BMO(P)}^2 \leq \frac{2(e^{\beta(C_p-1)} - 1)}{p(\beta(p-1) - p)},$$

since the right-hand side of (19) does not depends on τ.

$iii) \Longrightarrow iv)$ If M is a $BMO(P)$ martingale, then according to Lemma 2.1 it is sufficient to show that equation (10) admits bounded positive solution for some $p > 1$, which can be proved similarly to the implication $i) \Longrightarrow ii)$. By the same way one can show that for the mapping H

$$X_t = E\left[1 + \int_t^T \left[\frac{p}{2(p-1)^2}x_s - \frac{1}{p-1}\varphi_s\right]d\langle M\rangle_s \Big| \mathcal{F}_t\right],$$

where $-\int_0^t \Phi_s dM_s + L_t$ is the martingale part of X, the inequality (15) holds with

$$\alpha(p) = \frac{3p\|M\|_{BMO(P)}^2}{2(p-1)^2 - (9p-6)\|M\|_{BMO(P)}^2},$$

$$\beta(p) = \frac{6(p-1)}{2(p-1)^2 - (9p-6)\|M\|_{BMO(P)}^2},$$

where $\lim_{p\to\infty}\alpha(p) = \lim_{p\to\infty}\beta(p) = 0$. So if we take p large enough we obtain that the mapping H is a contraction.

$iv) \Longrightarrow i)$ The proof is similar to the proof of the implication $ii) \Longrightarrow iii)$ and we only give a brief sketch of the proof.

Since $\mathcal{E}(M)$ satisfies the (A_p) condition for some $p > 1$, according to Lemma 2.1 the process $X_t = E\left[\{\mathcal{E}_{t,T}(M)\}^{-\frac{1}{p-1}}\Big|\mathcal{F}_t\right]$ is a bounded positive solution of equation (10), which can be written in the following equivalent form

$$X_t = X_0 - \int_0^t \left[\frac{p}{2(p-1)^2}X_s - \frac{p}{p-1}\varphi_s\right]d\langle M\rangle_s - \int_0^t \varphi_s d\tilde{M}_s + L_t$$

in terms of $\tilde{P}$ martingale $\tilde{M} = \langle M\rangle - M$. Note that $\langle \tilde{M}\rangle = \langle M\rangle$ and L is also a local $\tilde{P}$ martingale orthogonal to $\tilde{M}$.

Applying the Itô formula for $e^{-\beta X_T} - e^{-\beta X_\tau}$, using successively the elementary inequality $\frac{\beta^2}{2}\varphi_s^2 - \frac{\beta p}{p-1}\varphi_s \geq -\frac{p^2}{2(p-1)^2}$, the convexity of the function $e^{-\beta x}$ and the two-sided inequality $1 \leq X_t \leq D_p$, similarly to the implication $ii) \Longrightarrow iii)$ we obtain the following estimate for the BMO norm of $\tilde{M}$

$$\|\tilde{M}\|_{BMO(\tilde{P})}^2 \leq \frac{2(p-1)^2}{p(\beta-p)}\left(e^{\beta(D_p-1)} - 1\right)$$

valid for any $\beta > p$, where D_p is a constant from Definition 2.2. $\square$

2.2 Girsanov's transformation of BMO martingales and BSDEs

Let M be a continuous local P-martingale such that $\mathcal{E}(M)$ is a uniformly integrable martingale and let $d\tilde{P} = \mathcal{E}_T(M)dP$. To each continuous local martingale X we associate the process $\tilde{X} = \langle X, M \rangle - X$, which is a local $\tilde{P}$-martingale according to Girsanov's theorem. We denote this map by $\varphi : \mathcal{L}(P) \to \mathcal{L}(\tilde{P})$, where $\mathcal{L}(P)$ and $\mathcal{L}(\tilde{P})$ are classes of P and $\tilde{P}$ local martingales.

Let consider the process

$$Y_t = E^{\tilde{P}}\big[\langle X \rangle_T - \langle X \rangle_t \big| \mathcal{F}_t\big] = E\big[\mathcal{E}_{t,T}(M)(\langle X \rangle_T - \langle X \rangle_t)\big| \mathcal{F}_t\big]. \tag{20}$$

Since $\langle \tilde{X} \rangle = \langle X \rangle$ under either probability measure, it is evident that

$$\|Y\|_\infty = \|\tilde{X}\|^2_{BMO(\tilde{P})}. \tag{21}$$

Let $M \in BMO(P)$. According to Theorem 2.1 condition (R_p) is satisfied for some $p > 1$. The (R_p) condition and conditional energy inequality (Kazamaki [21], page 29) imply that for any $X \in BMO(P)$ the process Y is bounded, i.e., φ maps $BMO(P)$ into $BMO(\tilde{P})$. Moreover, as proved by Kazamaki [20, 21], $BMO(P)$ and $BMO(\tilde{P})$ are isomorphic under the mapping ϕ and for all $X \in BMO(P)$ the inequality

$$\|\tilde{X}\|^2_{BMO(\tilde{P})} \leq C^2_K(\tilde{M}) \cdot \|X\|^2_{BMO(P)} \tag{22}$$

is valid, where

$$C^2_K(\tilde{M}) = 2p \cdot 2^{1/p} \sup_\tau \left\| E^{\tilde{P}}\Big[\{\mathcal{E}_{\tau,T}(\tilde{M})\}^{-\frac{1}{p-1}} \Big| \mathcal{F}_\tau \Big] \right\|^{(p-1)/p}_\infty, \tag{23}$$

and p is such that

$$\|\tilde{M}\|_{BMO(\tilde{P})} < \sqrt{2}(\sqrt{p} - 1). \tag{24}$$

Note that the similar inequality holds for the inverse mapping ϕ^{-1}.

Similarly to Lemma 2.1 one can show that for any $X \in BMO(P)$ the process Y (defined by (20)) is a positive bounded solution of the $BSDE$

$$\begin{cases} Y_t = Y_0 - \langle X \rangle_t - \int_0^t \varphi_s d\langle M \rangle_s + \int_0^t \varphi_s dM_s + L_t, \\ Y_T = 0. \end{cases} \tag{25}$$

Indeed, it is evident that $(Y_t + \langle X \rangle_t)\mathcal{E}_t(M)$ is a local martingale. Since $\mathcal{E}_t(M) > 0$ P-a.s. for all $t \in [0, T]$, the process Y will be a special semimartingale with the decomposition

$$Y_t = Y_0 + A_t + \int_0^t \varphi_s dM_s + N_t, \tag{26}$$

where A is a predictable process of bounded variation and N is a local martingale orthogonal to M.

By the Itô formula

$$(Y_t + \langle X \rangle_t)\mathcal{E}_t(M) = \int_0^t \mathcal{E}_s(M)\big[dA_s + d\langle X \rangle_s + \varphi_s d\langle M \rangle_s\big] + local\ martingale,$$

which implies that $A_t = -\langle X \rangle_t - \int_0^t \varphi_s d\langle M \rangle_s$. Therefore, it follows from (26) that Y satisfies equation (25).

Now we give an alternative proof of this assertion, which also improves the constant in the inequality (22).

Theorem 2.2. If $M \in BMO(P)$, then $\phi : X \to \tilde{X}$ is an isomorphism of $BMO(P)$ onto $BMO(\tilde{P})$. In particular, the inequality

$$\frac{1}{\left(1+\frac{\sqrt{2}}{2}\|M\|_{BMO(P)}\right)}\|X\|_{BMO(P)} \leq \|\tilde{X}\|_{BMO(\tilde{P})} \leq \left(1+\frac{\sqrt{2}}{2}\|\tilde{M}\|_{BMO(\tilde{P})}\right)\|X\|_{BMO(P)}. \quad (27)$$

is valid for any $X \in BMO(P)$.

Proof: Applying the Itô formula to $(Y_\tau + \varepsilon)^p - (Y_T + \varepsilon)^p$ (for $0 < p < 1$, $\varepsilon > 0$) and taking conditional expectations we obtain

$$(Y_\tau + \varepsilon)^p - \varepsilon^p = E\Big[\int_\tau^T p(Y_s + \varepsilon)^{p-1} d\langle X\rangle_s \Big|\mathcal{F}_\tau\Big] + \frac{p(1-p)}{2} E\Big[\int_\tau^T (Y_s + \varepsilon)^{p-2} d\langle L^c\rangle_s \Big|\mathcal{F}_\tau\Big] +$$

$$+ E\Big[\int_\tau^T \Big(\frac{p(1-p)}{2}(Y_s + \varepsilon)^{p-2}\varphi_s^2 + p(Y_s + \varepsilon)^{p-1}\varphi_s\Big) d\langle M\rangle_s \Big|\mathcal{F}_\tau\Big] -$$

$$- E\Big[\Sigma_{\tau < s \leq T}\big((Y_s + \varepsilon)^p - (Y_{s-} + \varepsilon)^p - p(Y_{s-} + \varepsilon)^{p-1}\Delta Y_s\big)\Big|\mathcal{F}_\tau\Big]. \quad (28)$$

Because $f(x) = x^p$ is concave for $p \in (0,1)$, the last term in (28) is positive. Therefore, using the inequality

$$\frac{p(1-p)}{2}(Y_s + \varepsilon)^{p-2}\varphi_s^2 + p(Y_s + \varepsilon)^{p-1}\varphi_s + \frac{p}{2(1-p)}(Y_s + \varepsilon)^p \geq 0$$

from (28) we obtain

$$(Y_\tau + \varepsilon)^p - \varepsilon^p \geq E\Big[\int_\tau^T p(Y_s + \varepsilon)^{p-1} d\langle X\rangle_s \Big|\mathcal{F}_\tau\Big] -$$

$$- \frac{p}{2(1-p)} E\Big[\int_\tau^T (Y_s + \varepsilon)^p d\langle M\rangle_s \Big|\mathcal{F}_\tau\Big]. \quad (29)$$

Since $0 < p < 1$

$$p(\|Y\|_\infty + \varepsilon)^{p-1} E\Big[\langle X\rangle_T - \langle X\rangle_\tau \Big|\mathcal{F}_\tau\Big] \leq E\Big[\int_\tau^T p(Y_s + \varepsilon)^{p-1} d\langle X\rangle_s \Big|\mathcal{F}_\tau\Big],$$

from (29) we have

$$p(\|Y\|_\infty + \varepsilon)^{p-1} E\Big[\langle X\rangle_T - \langle X\rangle_\tau \Big|\mathcal{F}_\tau\Big] \leq (Y_\tau + \varepsilon)^p - \varepsilon^p + \frac{p}{2(1-p)} E\Big[\int_\tau^T (Y_s + \varepsilon)^p d\langle M\rangle_s \Big|\mathcal{F}_\tau\Big]$$

and taking norms in the both sides of the latter inequality we obtain

$$p(\|Y\|_\infty + \varepsilon)^{p-1} \cdot \|X\|_{BMO(P)}^2 \leq (\|Y\|_\infty + \varepsilon)^p - \varepsilon^p + \frac{p}{2(1-p)}(\|Y\|_\infty + \varepsilon)^p \cdot \|M\|_{BMO(P)}^2.$$

Taking the limit when $\varepsilon \to 0$ we will have that for all $p \in (0,1)$

$$\|X\|_{BMO(P)}^2 \leq \Big(\frac{1}{p} + \frac{1}{2(1-p)}\|M\|_{BMO(P)}^2\Big) \cdot \|Y\|_\infty.$$

Therefore,
$$\|X\|^2_{BMO(P)} \leq \min_{p\in(0,1)} \left(\frac{1}{p} + \frac{1}{2(1-p)}\|M\|^2_{BMO(P)}\right) \cdot \|Y\|_\infty =$$
$$= \left(1 + \frac{\sqrt{2}}{2}\|M\|_{BMO(P)}\right)^2 \cdot \|Y\|_\infty, \tag{30}$$

since the minimum of the function $f(p) = \frac{1}{p} + \frac{1}{2(1-p)}\|M\|^2_{BMO(P)}$ is attained for $p^* = \sqrt{2}/(\sqrt{2} + \|M\|_{BMO(P)})$ and $f(p^*) = \left(1 + \frac{\sqrt{2}}{2}\|M\|_{BMO(P)}\right)^2$.

Thus, from (30) and (21) we obtain
$$\frac{1}{\left(1 + \frac{\sqrt{2}}{2}\|M\|_{BMO(P)}\right)}\|X\|_{BMO(P)} \leq \|\tilde{X}\|_{BMO(\tilde{P})}.$$

Now we can use inequality (30) for the Girsanov transform of $\tilde{X}$.

Since $dP/d\tilde{P} = \mathcal{E}_T^{-1}(M) = \mathcal{E}_T(\tilde{M})$, $\tilde{M}, \tilde{X} \in BMO(\tilde{P})$ and
$$\varphi(\tilde{X}) = \langle \tilde{X}, \tilde{M}\rangle - \tilde{X} = X,$$
from (30) we get the inverse inequality:
$$\|\tilde{X}\|_{BMO(\tilde{P})} \leq \left(1 + \frac{\sqrt{2}}{2}\|\tilde{M}\|_{BMO(\tilde{P})}\right)\|X\|_{BMO(P)}. \tag{31}$$

□

Let us compare the constant
$$C(\tilde{M}) = 1 + \frac{\sqrt{2}}{2}\|\tilde{M}\|_{BMO(\tilde{P})}$$
from (27) with the corresponding constant $C_K(\tilde{M})$ from (22) (Kazamaki [20, 21]). Since
$$E^{\tilde{P}}\left[\{\mathcal{E}_{\tau,T}(\tilde{M})\}^{-\frac{1}{p-1}}\Big|\mathcal{F}_\tau\right] \geq 1,$$
the constant $C_K(\tilde{M})$ is more than $\sqrt{2p}$, where p is such that $\|\tilde{M}\|_{BMO(\tilde{P})} < \sqrt{2}(\sqrt{p}-1)$. Since the last inequality is equivalent to the inequality $p > \left(1 + \frac{\sqrt{2}}{2}\|\tilde{M}\|_{BMO(\tilde{P})}\right)^2$, we obtain that at least
$$C^2(\tilde{M}) \leq \frac{1}{2}C_K^2(\tilde{M}).$$

It is evident that in the trivial case $M = 0$, we have that $\tilde{P} = P$ and $\tilde{X} = X$. Note that, if $M = 0$ then (27) gives the two-sided inequality
$$\|X\|_{BMO(P)} \leq \|\tilde{X}\|_{BMO(\tilde{P})} \leq \|X\|_{BMO(P)},$$
implying the equality $\tilde{X} = X$, whereas from (22), we only have
$$\frac{1}{2}\|X\|_{BMO(P)} \leq \|\tilde{X}\|_{BMO(\tilde{P})} \leq 2\|X\|_{BMO(P)}.$$

This shows that the following simple corollary cannot be deduced from inequality (22).

Corollary. Let $(M^n, n \geq 1)$ be a sequence of $BMO(P)$ martingales with $\lim_{n\to\infty} ||M^n||_{BMO(P)} = 0$. Let $dP^n = \mathcal{E}_T(M^n)dP$ and $\tilde{X}^n = \langle X, M^n \rangle - X$. Then for any $X \in BMO(P)$

$$\lim_{n\to\infty} ||\tilde{X}^n||_{BMO(P^n)} = ||X||_{BMO(P)}.$$

Proof. The second inequality of (27), applied for $X = M^n$ and $M = M^n$ gives

$$||\tilde{M}^n||_{BMO(P^n)} \leq \left(1 + \frac{\sqrt{2}}{2}||\tilde{M}^n||_{BMO(P^n)}\right)||M^n||_{BMO(P)}.$$

Therefore,

$$\frac{1}{\frac{\sqrt{2}}{2} + 1/||\tilde{M}^n||_{BMO(P^n)}} \leq ||M^n||_{BMO(P)},$$

which implies that $\lim_{n\to\infty} ||\tilde{M}^n||_{BMO(P^n)} = 0$. Now, passing to the limit in the two-sided inequality (27) we obtain

$$||X||_{BMO(P)} \leq \lim_{n\to\infty} ||\tilde{X}^n||_{BMO(P^n)} \leq ||X||_{BMO(P)}. \quad \square$$

Remark. Note that the converse of Theorem 2.2 is also true. I.e., if M is a continuous local martingale and $\mathcal{E}(M)$ is a uniformly integrable martingale, Schachermayer [34] proved that if $M \notin BMO(P)$ then the map φ is not an isomorphism from $BMO(P)$ into $BMO(\tilde{P})$.

3 Backward stochastic differential equations with a convex generator

As it was mentioned above in this chapter our aim is to prove the existence and uniqueness result for BSDE with convex generator of quadratic growth. The main theorem is the following:

Theorem 3.1. Suppose that the filtration $\{\mathcal{F}_t\}_{0\leq t\leq T}$ is continuous and $\{M_t\}_{0\leq t\leq T}$ is a martingale from the class BMO. Let the parameters (f, η) of equation (1) satisfy the following conditions:
1) for any (t, ω), $f(t, \omega, \cdot)$ is a continuous and convex function.
2) there exist a predictable non-negative process α_t and a constant $\gamma \geq 0$ such that $\int \alpha dM \in BMO$ and for any (t, ω, z)
$$|f(t,\omega,z)| \leq \alpha_t(\omega) + \frac{\gamma}{2}z^2.$$

3) $E e^{\gamma \eta + \gamma \int_0^T \alpha_s d\langle M \rangle_s} < \infty$ and $\eta + \int_0^T f(s,0)d\langle M \rangle_s \geq -C$ for some $C \geq 0$.

Then there exists a solution $V = \{V_t\}_{0\leq t\leq T}$ of equation (1) represented in the form

$$V_t = \operatorname{ess\,sup}_{u \in U} E\left[\mathcal{E}_{t,T}\left(\int f'_l(u)dM\right) \left(\eta + \int_t^T [f(s,u_s) - f'_l(s,u_s)u_s]d\langle M \rangle_s\right) \Big| \mathcal{F}_t \right]$$

where U is the class of predictable bounded controls.

We prove **Theorem 3.1** in four sections of this chapter. In the first section we show that the value process V is a supersolution of equation (1); In the second we consider the case when the main parameters of equation (1) are bounded; In the third section we prove **Theorem 3.1** when the generator and boundary condition are nonnegative and finally we finish the proof of **Theorem 3.1** in the forth section. After that, in the fifth section we provide the multi dimensional analogue of **Theorem 3.1** without the proof, because it is similar to the proof of **Theorem 3.1**. In the next section under some additional conditions, we obtain that equation (1) admits unique solution in special class of solutions. At the end of this chapter we provide an appendix for some auxiliary assertions.

3.1 Value process as a supersolution of equation (1)

For any $u \in U$, let us consider the linear $(BSDE)$

$$\begin{cases} Y_t = Y_0 - \int_0^t [f(s,u_s) + f'_l(s,u_s)(Z_s - u_s)]d\langle M \rangle_s + \int_0^t Z_s dM_s + L_t, & (6) \\ Y_T = \eta. \end{cases}$$

Now, according to condition **2)** of **Theorem 3.1** and from Lemma 3.14 (see Appendix) it follows that for any $u \in U$, $\mathcal{E}_t(\int f'_l(u)dM)$ is a uniformly integrable martingale. This means that for any $u \in U$, $dP^u = \mathcal{E}_T(\int f'_l(u)dM)dP$ is a probability measure, where $\mathcal{E}_t(\int f'_l(u)dM)$ is the stochastic exponent from the martingale $\int f'_l(u)dM$.

Thus there exists the unique solution of equation (6):

$$Y^u_t = E^u\left[\eta + \int_t^T [f(s,u_s) - f'_l(s,u_s)u_s]d\langle M \rangle_s \Big| \mathcal{F}_t \right].$$

Let us consider the process $\{V_t\}_{0\leq t\leq T}$:

$$V_t = \operatorname*{ess\,sup}_{u\in U} Y_t^u = \operatorname*{ess\,sup}_{u\in U} E^u\left[\eta + \int_t^T [f(s,u_s) - f'_l(s,u_s)u_s]d\langle M\rangle_s \bigg| \mathcal{F}_t\right].$$

From this representation it is obvious that $V_T = \eta$.

Definition 3.1. We say that the process $(Y_t)_{0\leq t\leq T}$ is a supersolution of equation (1) if there exists an increasing process $(K_t)_{0\leq t\leq T}$ such that Y_t satisfies the equation

$$\begin{cases} Y_t = Y_0 - \int_0^t f(s, Z_s)d\langle M\rangle_s - K_t + \int_0^t Z_s dM_s + L_t, \\ Y_T = \eta, \end{cases}$$

where Z is a predictable process for which $\int_0^T Z_s^2 d\langle M\rangle_s < \infty$ P a.s. and L is a local martingale orthogonal to M.

Our aim is to show that V satisfies equation (1). In this section we shall only show that V is a supersolution of equation (1). For this we need the optimality principle which is proved in a standard manner [17] and in our case takes the following form.

Proposition 3.1 (optimality principle). *There exists a right continuous left-limited (RCLL) modification of the process V_t such that:*
  a) *for any $u \in U$ the process $V_t + \int_0^t [f(s,u_s) - f'_l(s,u_s)u_s]d\langle M\rangle_s$ is a supermartingale with respect to measure P^u.*
  b) *u^* is optimal if and only if $V_t + \int_0^t [f(s,u_s^*) - f'_l(s,u_s^*)u_s^*]d\langle M\rangle_s$ is a martingale with respect to the measure P^{u^*}.*

Proposition 3.2. *Suppose $\{M_t\}_{0\leq t\leq T}$ is a martingale from the class BMO and let the parameters (f,η) of equation (1) satisfy the following conditions:*
  1) *for any (t,ω), $f(t,\omega,\cdot)$ is a continuous and convex function.*
  2) *there exists a predictable non-negative process α_t and a constant $\gamma \geq 0$ such that $\int \alpha dM \in BMO$ and for any (t,ω,z)*

$$|f(t,\omega,z)| \leq \alpha_t(\omega) + \frac{\gamma}{2}z^2.$$

Then the process $(V_t)_{0\leq t\leq T}$ is a supersolution of equation (1).

Proof. According to Proposition 3.1, $V_t + \int_0^t [f(s,u_s) - f'_l(s,u_s)u_s]d\langle M\rangle_s$ is a P^u-supermartingale. Therefore the same process will be a semimartingale with respect to the measure P. This implies that the process V_t is a P-semimartingale. So we have the semimartingale decomposition of V:

$$V_t = V_0 + A_t + \int_0^t \varphi_s dM_s + L_t,$$

where φ is a predictable process with $\int_0^T \varphi_s^2 d\langle M\rangle_s < \infty$ P a.s, L is a local martingale orthogonal to M and A is a process of bounded variation. Here we use the Kunita–Watanabe decomposition, which says that every locally square integrable martingale is represented as a sum of stochastic integral (with respect to M) and a local martingale orthogonal to M ([24]). Now using this decomposition we have

$$V_t + \int_0^t [f(s,u_s) - f_l'(s,u_s)u_s]d\langle M\rangle_s = V_0 + A_t + \int_0^t [f(s,u_s) - f_l'(s,u_s)u_s]d\langle M\rangle_s + \int_0^t \varphi_s dM_s + L_t =$$

$$= V_0 + A_t + \int_0^t [f(s,u_s) - f_l'(s,u_s)u_s]d\langle M\rangle_s + \int_0^t \varphi_s f_l'(s,u_s)d\langle M\rangle_s +$$

$$+ \int_0^t \varphi_s dM_s - \int_0^t \varphi_s f_l'(s,u_s)d\langle M\rangle_s + L_t =$$

$$= V_0 + \int_0^t \varphi_s dM_s - \int_0^t \varphi_s f_l'(s,u_s)d\langle M\rangle_s + L_t + A_t + \int_0^t [f(s,u_s) + f_l'(s,u_s)(\varphi_s - u_s)]d\langle M\rangle_s.$$

According to the Girsanov's theorem, $\int \varphi dM - \int \varphi f_l'(u)d\langle M\rangle$ and L are orthogonal P^u-local martingales. Thus $A_t + \int_0^t [f(s,u_s) + f_l'(s,u_s)(\varphi_s - u_s)]d\langle M\rangle_s$ is a decreasing process. Since A_t does not depend on u, the process

$$A_t + \operatorname{ess\,sup}_{u\in U} \int_0^t [f(s,u_s) + f_l'(s,u_s)(\varphi_s - u_s)]d\langle M\rangle_s$$

is also decreasing.

Now let us show that

$$\operatorname{ess\,sup}_{u\in U} \int_0^t [f(s,u_s) + f_l'(s,u_s)(\varphi_s - u_s)]d\langle M\rangle_s = \int_0^t f(s,\varphi_s)d\langle M\rangle_s.$$

According to Lemma 3.15 (see Appendix),

$$\operatorname{ess\,sup}_{u\in U_n}[f(s,u_s) + f_l'(s,u_s)(\varphi_s - u_s)] = f(s,\varphi_s^n) + f_l'(s,\varphi_s^n)(\varphi_s - \varphi_s^n) = g^n(s,\varphi_s),$$

where $U_n = \{u \in U : |u_s(\omega)| \leq n\}$ and the function g^n has the form

$$g^n(s,x) = \mathbf{1}_{(x>n)}[f(s,n) + f_l'(s,n)(x-n)] + \mathbf{1}_{(|x|\leq n)}f(s,x) + \mathbf{1}_{(x<-n)}[f(s,-n) + f_l'(s,-n)(x+n)].$$

This equality shows that in the class of controls U_n the sup is attained for the control φ^n where $\varphi^n = n\mathbf{1}_{(\varphi>n)} + \varphi\mathbf{1}_{(|\varphi|\leq n)} - n\mathbf{1}_{(\varphi<-n)}$. From this fact we have the equality

$$\operatorname{ess\,sup}_{u\in U_n} \int_0^t [f(s,u_s) + f_l'(s,u_s)(\varphi_s - u_s)]d\langle M\rangle_s = \int_0^t \operatorname{ess\,sup}_{u\in U_n} [f(s,u_s) + f_l'(s,u_s)(\varphi_s - u_s)]d\langle M\rangle_s.$$

With this we have the convergence $g^n(s,\varphi_s) \uparrow f(s,\varphi_s)$ P a.s. and for any n, $g^n(s,\varphi_s) \geq g^0(s,\varphi_s) = f(s,0) + f_l'(s,0)\varphi_s$. Now P a.s.

$$\int_0^t g^0(s,\varphi_s)d\langle M\rangle_s = \int_0^t f(s,0)d\langle M\rangle_s + \int_0^t f_l'(s,0)\varphi_s d\langle M\rangle_s \geq$$

$$\geq -\int_0^t \alpha_s d\langle M\rangle_s - \int_0^t (\gamma + 2\alpha_s)|\varphi_s|d\langle M\rangle_s \geq$$

$$\geq -\int_0^t \alpha_s d\langle M\rangle_s - \sqrt{\int_0^t (\gamma + 2\alpha_s)^2 d\langle M\rangle_s} \cdot \sqrt{\int_0^t |\varphi_s|^2 d\langle M\rangle_s} > -\infty.$$

Therefore, by the monotonic convergence theorem we have that P a.s.

$$\int_0^t g^n(s,\varphi_s)d\langle M\rangle_s \longrightarrow \int_0^t f(s,\varphi_s)d\langle M\rangle_s.$$

Now according to the above-mentioned facts we obtain the following chain of equalities:

$$\operatorname{ess\,sup}_{u\in U}\int_0^t [f(s,u_s)+f'_l(s,u_s)(\varphi_s-u_s)]d\langle M\rangle_s$$
$$=\sup_n\left\{\operatorname{ess\,sup}_{u\in U_n}\int_0^t [f(s,u_s)+f'_l(s,u_s)(\varphi_s-u_s)]d\langle M\rangle_s\right\}$$
$$=\lim_{n\to\infty}\operatorname{ess\,sup}_{u\in U_n}\int_0^t [f(s,u_s)+f'_l(s,u_s)(\varphi_s-u_s)]d\langle M\rangle_s$$
$$=\lim_{n\to\infty}\int_0^t \operatorname{ess\,sup}_{u\in U_n}[f(s,u_s)+f'_l(s,u_s)(\varphi_s-u_s)]d\langle M\rangle_s$$
$$=\lim_{n\to\infty}\int_0^t g^n(s,\varphi_s)d\langle M\rangle_s = \int_0^t \lim_{n\to\infty}g^n(s,\varphi_s)d\langle M\rangle_s = \int_0^t f(s,\varphi_s)d\langle M\rangle_s.$$

From this we obtain that $A_t + \int_0^t f(s,\varphi_s)d\langle M\rangle_s$ is a decreasing process. Now consider the process $-K_t = A_t + \int_0^t f(s,\varphi_s)d\langle M\rangle_s$. It is obvious that K_t is an increasing process. Now if we insert this equality in the semimartingale decomposition of V, we will have

$$V_t = V_0 - \int_0^t f(s,\varphi_s)d\langle M\rangle_s - K_t + \int_0^t \varphi_s dM_s + L_t. \qquad \square$$

3.2 An existence of a solution in the case of bounded characteristics

In this section we assume that $\|\eta\|_\infty + \|\langle M\rangle_T\|_\infty + \|\int_0^T \alpha_t^2 d\langle M\rangle_t\|_\infty < \infty$ and in this case we prove that the value process V is a bounded solution of equation (1).

Proposition 3.3. *The process V is bounded.*

Proof. Let η, $\langle M\rangle_T$ and $\int_0^T \alpha_t^2 d\langle M\rangle_t$ be bounded with a constant D. Then since f is convex $f(s,u_s) - f'_l(s,u_s)u_s \leq f(s,0)$ and according to condition **2)** of Theorem 3.1, $|f(s,0)| \leq \alpha_s$. Now if we use these two inequalities, we obtain

$$V_t = \operatorname{ess\,sup}_{u\in U} E^u\left[\eta + \int_t^T [f(s,u_s)-f'_l(s,u_s)u_s]d\langle M\rangle_s\Big|\mathcal{F}_t\right]$$
$$\leq \operatorname{ess\,sup}_{u\in U} E^u\left[\|\eta\|_\infty + \int_t^T f(s,0)d\langle M\rangle_s\Big|\mathcal{F}_t\right]$$
$$\leq \operatorname{ess\,sup}_{u\in U} E^u\left[\|\eta\|_\infty + \int_t^T \alpha_s d\langle M\rangle_s\Big|\mathcal{F}_t\right]$$
$$\leq \|\eta\|_\infty + \left\|\int_0^T \alpha_s d\langle M\rangle_s\right\|_\infty \leq 2D.$$

On the other hand, for the control $u^0 \equiv 0$ we have

$$V_t \geq E^{u^0}\left[\eta + \int_t^T [f(s, u_s^0) - f_l'(s, u_s^0)u_s^0]d\langle M\rangle_s \Big| \mathcal{F}_t\right]$$

$$\geq -\|\eta\|_\infty + E^{u^0}\left[\int_t^T f(s, 0)d\langle M\rangle_s \Big| \mathcal{F}_t\right]$$

$$\geq -\|\eta\|_\infty - E^{u^0}\left[\int_t^T \alpha_s d\langle M\rangle_s \Big| \mathcal{F}_t\right]$$

$$\geq -\|\eta\|_\infty - \left\|\int_0^T \alpha_s d\langle M\rangle_s\right\|_\infty \geq -2D > -\infty.$$

This means that V is bounded with the constant $2D$. □

Now for any $n \in N$ let us consider the following process:

$$V_t^n = \operatorname{ess\,sup}_{u \in U_n} E^u\left[\eta + \int_t^T [f(s, u_s) - f_l'(s, u_s)u_s]d\langle M\rangle_s \Big| \mathcal{F}_t\right],$$

where U_n is the class of predictable controls bounded with n. It is evident that for any $t \in [0; T]$ $V_t^n \leq V_t$ and P a.s. $V_t = \uparrow \lim_{n\to\infty} V_t^n$, but we shall show that the probability of such ω that $\lim_{n\to\infty} V_t^n(\omega) = V_t(\omega)$ for any $t \in [0; T]$, is equal to 1.

Lemma 3.1. *For the processes V and V^n the following equality holds:*

$$P\left(\omega : \lim_{n\to\infty} V_t^n(\omega) = V_t(\omega) \ \forall \ t \in [0; T]\right) = 1.$$

Proof. As we have mentioned above, for any $t \in [0; T]$, $V_t = \sup_n V_t^n = \lim_{n\to\infty} V_t^n$ P a.s.

For any $n \in N$, consider the process $X_t^n = V_t^n + \int_0^t f(s, 0)d\langle M\rangle_s$. Let P^0 be the measure corresponding to $u^0 \equiv 0$. It is evident that $u^0 \in U_n$ for any $n \in N$. According to Proposition 3.1, $X_t^n = V_t^n + \int_0^t f(s, 0)d\langle M\rangle_s$ and $Y_t = V_t + \int_0^t f(s, 0)d\langle M\rangle_s$ are right continuous supermartingales with respect to the measure P^0. So, as a result we have an increasing, right continuous sequence of P^0-supermartingales $(X^n)_{n\geq 1}$. With this for any $t \in [0; T]$ P a.s. we have

$$Y_t = V_t + \int_0^t f(s, 0)d\langle M\rangle_s = \lim_{n\to\infty}\left[V_t^n + \int_0^t f(s, 0)d\langle M\rangle_s\right]$$

$$= \sup_n\left[V_t^n + \int_0^t f(s, 0)d\langle M\rangle_s\right] = \sup_n X_t^n.$$

According to Theorem T16 [14], since $\sup_n X^n$ is right continuous we obtain that

$$P^0\left(V_t + \int_0^t f(s, 0)d\langle M\rangle_s = \sup_n\left[V_t^n + \int_0^t f(s, 0)d\langle M\rangle_s\right] \ \forall \ t \in [0; T]\right) = 1.$$

The following equality $\sup_n\left[V_t^n + \int_0^t f(s, 0)d\langle M\rangle_s\right] = \sup_n V_t^n + \int_0^t f(s, 0)d\langle M\rangle_s$, implies that $V_t = \sup_n V_t^n = \lim_{n\to\infty} V_t^n$ and because P and P^0 are equivalent measures,

$$P\left(\omega : \lim_{n\to\infty} V_t^n(\omega) = V_t(\omega) \ \forall \ t \in [0; T]\right) = 1. \qquad \Box$$

Now we shall write the equation for the process V^n:

Proposition 3.4. V^n satisfies the following backward stochastic differential equation:
$$\begin{cases} Y_t = Y_0 - \int_0^t g^n(s, Z_s) d\langle M \rangle_s + \int_0^t Z_s dM_s + L_t, \\ Y_T = \eta, \end{cases} \quad (32)$$
where g^n is the function from Lemma 3.15.

Proof. According to Proposition 3.1, for any $u \in U_n$
$$V_t^n + \int_0^t [f(s, u_s) - f_l'(s, u_s) u_s] d\langle M \rangle_s$$
is a P^u-supermartingale. So the process V_t^n is a P-semimartingale and we have the semimartingale decomposition of V^n
$$V_t^n = V_0^n + A_t^n + \int_0^t \varphi_s^n dM_s + L_t^n,$$
where φ^n is a predictable process for which $\int_0^T (\varphi^n)_s^2 d\langle M \rangle_s < \infty$ P a.s. L^n is a local martingale orthogonal to M and A^n is a process of bounded variation. In the same way as we have done for the process V, we obtain that the process

$$A_t^n + \operatorname{ess\,sup}_{u \in U_n} \int_0^t [f(s, u_s) + f_l'(s, u_s)(\varphi_s^n - u_s)] d\langle M \rangle_s$$
$$= A_t^n + \int_0^t \operatorname{ess\,sup}_{u \in U_n} [f(s, u_s) + f_l'(s, u_s)(\varphi_s^n - u_s)] d\langle M \rangle_s$$
$$= A_t^n + \int_0^t g^n(s, \varphi_s^n) d\langle M \rangle_s$$

is a decreasing process. Since the space of values of controls from the class U_n is a compact set, there exists an optimal control $u^* \in U_n$ ([16] and [10]). So, according to part **2)** of Proposition 3.1,
$$V_t^n + \int_0^t [f(s, u_s^*) - f_l'(s, u_s^*) u_s^*] d\langle M \rangle_s$$
is a P^{u^*}-martingale. This implies that
$$A_t^n + \int_0^t [f(s, u_s^*) + f_l'(s, u_s^*)(\varphi_s^n - u_s^*)] d\langle M \rangle_s = 0.$$

Now using these facts we obtain
$$A_t^n = -\int_0^t \operatorname{ess\,sup}_{u \in U_n} [f(s, u_s) + f_l'(s, u_s)(\varphi_s^n - u_s)] d\langle M \rangle_s$$
$$= -\int_0^t g^n(s, \varphi_s^n) d\langle M \rangle_s.$$

This means that V^n satisfies the following equation
$$\begin{cases} V_t^n = V_0^n - \int_0^t g^n(s, \varphi_s^n) d\langle M \rangle_s + \int_0^t \varphi_s^n dM_s + L_t^n, \\ V_T^n = \eta. \end{cases} \quad (33)$$
$\square$

Now we are ready to prove that when $\|\eta\|_\infty + \|\langle M\rangle_T\|_\infty + \|\int_0^T \alpha_t^2 d\langle M\rangle_t\|_\infty < \infty$, the value process V is a solution of equation (1). For this we need the Monotone Stability Lemma which was proved by Kobylanski [22] in the Brownian setting and then was generalized by A. Morlais [31] for continuous martingales. This Lemma in our case has the following form.

Lemma 3.2. *(Monotone Stability). Considering equation (1) given by its parameters (f, η), we give here the properties required on the two sequences f^n and η^n associated with the BSDEs given by (f^n, η^n):*

1) *P-a.s. and for all s, the sequence $(f^n : z \to f^n(z))$ converges increasingly w.r.t n and uniformly on the compact sets of R to $(f : z \to f(z))$ (f is continuous w.r.t z)*

2) *For all n, f^n satisfies the quadratic growth condition with the same parameters $\bar{\alpha}_s, \bar{\gamma}$: $|f^n(s,x)| \leq \bar{\alpha}_s + \bar{\gamma} x^2$ where $\|\int_0^T \bar{\alpha}_t d\langle M\rangle_t\|_\infty < \infty$.*

3) *(η^n) is a uniformly bounded sequence of $\mathcal{F}_T$-measurable random variables, which converges increasingly w.r.t n and P-a.s. to η.*

If there exists a solution (Y^n, Z^n, L^n) of the BSDEs given by (f^n, η^n) such that the sequence $(Y^n)_n$ is increasing, then the sequence (Y^n, Z^n, L^n) converges to $(\tilde{Y}, \tilde{Z}, \tilde{L})$ in the following sense

$$E\left(\sup_{t\in[0;T]} |Y_t^n - \tilde{Y}_t|\right) \longrightarrow 0 \quad as \quad n \to \infty,$$

$$E\left(\int_0^T |\tilde{Z}_s - Z_s^n|^2 d\langle M\rangle_s + |\tilde{L}_T - L_T^n|^2\right) \longrightarrow 0 \quad as \quad n \to \infty.$$

Besides, $(\tilde{Y}, \tilde{Z}, \tilde{L})$ is a solution of the BSDE of equation (1) with parameters (f, η).

Let us verify that the conditions of this lemma hold.

In our case the parameters are (g^n, η). As we know from the definition of g^n, $(g^n)_n$ is increasing sequence of generators and converges to f. Moreover, since g^n and f are continuous, the convergence is uniform on the compact subsets of R. Here the terminal condition η is bounded and does not depend on n. According to Proposition 3.4, the triple (V^n, φ^n, L^n) is a solution of the backward equation with parameters (g^n, η) and is increasing w.r.t n. So we need only to verify the second condition of the Monotone stability Lemma. For this we need to find a nonnegative process $\bar{\alpha}$ and a constant $\bar{\gamma}$ such that the following inequality holds for any $n \in N$: $|g^n(s,x)| \leq \bar{\alpha}_s + \bar{\gamma} x^2$.

From the definition of g^n it follows that $g^n(s,x) \leq f(s,x) \leq \alpha_s + \frac{\gamma}{2} x^2$ and for the lower bound we have

$$g^n(s,x) \geq g^0(s,x) = f(s,0) + f'_l(s,0)x \geq -\alpha_s - |f'_l(s,0)| \cdot |x|$$

$$\geq -\alpha_s - \frac{1}{2}(|f'_l(s,0)|^2 + |x|^2) \geq -\alpha_s - \frac{1}{2}|2\alpha_s + \gamma|^2 - \frac{1}{2}|x|^2$$

$$\geq -\gamma^2 - \alpha_s - 4\alpha_s^2 - \frac{1}{2}|x|^2.$$

If we take $\bar{\alpha}_s = \gamma^2 + \alpha_s + 4\alpha_s^2$ and $\bar{\gamma} = \frac{\gamma+1}{2}$, then we obtain the estimation of g^n: $|g^n(s,x)| \leq \bar{\alpha}_s + \bar{\gamma} x^2$.

Finally, since $\langle M\rangle_T$ and $\int_0^T \alpha_t^2 d\langle M\rangle_t$ are bounded, we obtain that $\int_0^T \bar{\alpha}_t d\langle M\rangle_t$ is bounded too.

Now we are ready to use the Monotone Stability Lemma. As a result we obtain that there exists a triple $(\tilde{V}, \varphi, L)$ such that

$$E\left(\sup_{t\in[0;T]} |V_t^n - \tilde{V}_t|\right) \longrightarrow 0 \quad as \quad n \to \infty,$$

$$E\left(\int_0^T |\varphi_s - \varphi_s^n|^2 d\langle M\rangle_s + |L_T - L_T^n|^2\right) \longrightarrow 0 \quad as \quad n \to \infty.$$

If we take limits to the both sides of equation (33), we obtain that the triple $(\tilde{V}, \varphi, L)$ is a solution of equation (1). But according to Lemma 3.1, V and $\tilde{V}$ are indistinguishable. This means that the value process V is a bounded solution of equation (1).

3.3 The case where the generator f and η are nonnegative

In this section we prove Theorem 3.1 where the generator f and η are nonnegative. Before that let us prove a lemma which gives an a priori estimate for the solutions of equation (1).

Lemma 3.3. *If $\|\eta\|_\infty + \|\int_0^T \alpha_s d\langle M\rangle_s\|_\infty < \infty$, then for any bounded solution of equation (1) the following inequality holds:*

$$Y_t \leq \frac{1}{\gamma} \ln E\left[e^{\gamma\eta + \gamma\int_t^T \alpha_s d\langle M\rangle_s} \Big| \mathcal{F}_t\right].$$

Proof. By the Itô formula,

$$e^{\gamma Y_t + \gamma \int_0^t \alpha_s d\langle M\rangle_s} = e^{\gamma Y_0} - \gamma \int_0^t e^{\gamma Y_s + \gamma \int_0^s \alpha_r d\langle M\rangle_r} f(s, Z_s) d\langle M\rangle_s$$
$$+ \gamma \int_0^t e^{\gamma Y_s + \gamma \int_0^s \alpha_r d\langle M\rangle_r} Z_s dM_s + \gamma \int_0^t e^{\gamma Y_s + \gamma \int_0^s \alpha_r d\langle M\rangle_r} dL_s$$
$$+ \gamma \int_0^t e^{\gamma Y_s + \gamma \int_0^s \alpha_r d\langle M\rangle_r} \alpha_s d\langle M\rangle_s + \frac{\gamma^2}{2} \int_0^t e^{\gamma Y_s + \gamma \int_0^s \alpha_r d\langle M\rangle_r} Z_s^2 d\langle M\rangle_s$$
$$+ \frac{\gamma^2}{2} \int_0^t e^{\gamma Y_s + \gamma \int_0^s \alpha_r d\langle M\rangle_r} d\langle L\rangle_s. \quad (*)$$

Since $f(s, Z_s) \leq \alpha_s + \frac{\gamma}{2} Z_s^2$, we obtain that the process

$$-\gamma \int_0^t e^{\gamma Y_s + \gamma \int_0^s \alpha_r d\langle M\rangle_r} f(s, Z_s) d\langle M\rangle_s + \gamma \int_0^t e^{\gamma Y_s + \gamma \int_0^s \alpha_r d\langle M\rangle_r} \alpha_s d\langle M\rangle_s$$
$$+ \frac{\gamma^2}{2} \int_0^t e^{\gamma Y_s + \gamma \int_0^s \alpha_r d\langle M\rangle_r} Z_s^2 d\langle M\rangle_s$$
$$= \gamma \int_0^t e^{\gamma Y_s + \gamma \int_0^s \alpha_r d\langle M\rangle_r} \left(\frac{\gamma}{2} Z_s^2 - f(s, Z_s) + \alpha_s\right) d\langle M\rangle_s$$

is an increasing process. Since the stochastic integrals in $(*)$ are martingales, this means that $e^{\gamma Y_t + \gamma \int_0^t \alpha_s d\langle M\rangle_s}$ is a submartingale. So we have the following inequality

$$e^{\gamma Y_t + \gamma \int_0^t \alpha_s d\langle M\rangle_s} \leq E\left[e^{\gamma\eta + \gamma\int_0^T \alpha_s d\langle M\rangle_s} \Big| \mathcal{F}_t\right].$$

From this it is clear that

$$Y_t \leq \frac{1}{\gamma} \ln E\left[e^{\gamma\eta + \gamma\int_t^T \alpha_s d\langle M\rangle_s} \Big| \mathcal{F}_t\right]. \qquad \square$$

Proposition 3.5. *Let the conditions of Theorem 3.1 be satisfied, and η and f be nonnegative. Then there exists a solution (Y, φ, L) of equation (1), where Y is represented in the form*

$$Y_t = \operatorname*{ess\,sup}_{u \in U} E^u\left[\eta + \int_t^T [f(s, u_s) - f_l'(s, u_s)u_s] d\langle M\rangle_s \Big| \mathcal{F}_t\right].$$

Proof. Let $\eta^n = \eta \wedge n$ and define a sequence of stopping times

$$\tau_n = \inf\left\{t \geq 0 : \int_0^t \alpha_s^2 d\langle M\rangle_s \vee \langle M\rangle_t \geq n\right\}.$$

Let us consider the equation

$$\begin{cases} Y_t = Y_0 - \int_0^t 1_{(s\leq \tau_n)} f(s, Z_s) d\langle M\rangle_s + \int_0^t Z_s dM_s + L_t, \\ Y_T = \eta^n. \end{cases} \quad (34)$$

To solve equation (34) for any n, we need the result of **Section 3.2**. For this one should verify condition **2)** of the Monotone Stability Lemma for the generators $(1_{(s\leq \tau_n)} g^m(s,x))_{m\geq 1}$, where the function g^m is defined in Lemma 3.15 (see Appendix). For a fixed n, taking $\bar{\alpha}_s^n = 1_{(s\leq \tau_n)}(\gamma^2 + \alpha_s + 4\alpha_s^2)$ and $\bar{\gamma} = \frac{1+\gamma}{2}$ it is clear that we obtain an estimate

$$|1_{(s\leq \tau_n)} g^m(s,x)| \leq \bar{\alpha}_s^n + \bar{\gamma} x^2.$$

Now, if we recall the definition of τ_n we obtain that $\int_0^T \bar{\alpha}_s^n d\langle M\rangle_s$ is bounded. According to **Section 3.2**, this means that for any n there exists a bounded solution (Y^n, φ^n, L^n) of equation (34)

$$\begin{cases} Y_t^n = Y_0^n - \int_0^t 1_{(s\leq \tau_n)} f(s, \varphi_s^n) d\langle M\rangle_s + \int_0^t \varphi_s^n dM_s + L_t^n, \\ Y_T^n = \eta^n, \end{cases} \quad (35)$$

where Y^n has the form

$$Y_t^n = \operatorname{ess\,sup}_{u\in U} E\left[\mathcal{E}_{t,T}\left(\int 1_{(s\leq \tau_n)} f_l'(u) dM\right)\left(\eta^n + \int_t^T 1_{(s\leq \tau_n)}[f(s,u_s) - f_l'(s,u_s)u_s] d\langle M\rangle_s\right)\bigg|\mathcal{F}_t\right].$$

Easy calculations show that

$$Y_t^n = \operatorname{ess\,sup}_{u\in U} E\left[\mathcal{E}_{t\wedge \tau_n, \tau_n}\left(\int f_l'(u) dM\right)\left(\eta^n + \int_{t\wedge \tau_n}^{\tau_n}[f(s,u_s) - f_l'(s,u_s)u_s] d\langle M\rangle_s\right)\bigg|\mathcal{F}_t\right].$$

With this in view let us define the value process

$$V_t = \operatorname{ess\,sup}_{u\in U} E^u\left[\eta + \int_t^T [f(s,u_s) - f_l'(s,u_s)u_s] d\langle M\rangle_s \bigg|\mathcal{F}_t\right].$$

At this moment our aim is to show that the sequence Y^n is increasing and $V_t = \sup_n Y_t^n$.

Lemma 3.4. *For any $n \in \mathbb{N}$, $Y^n \leq Y^{n+1}$.*

Proof. Note that since f is nonnegative and Y^n is the bounded solution of equation (34), it follows that for any n, Y^n is a supermartingale.

For simplicity, let us introduce some notation:

$$g(s, u_s) := f(s, u_s) - f_l'(s, u_s)u_s,$$

$$H_t^n(u) := E\left[\mathcal{E}_{t\wedge \tau_n, \tau_n}\left(\int f_l'(u) dM\right)\left(\eta^n + \int_{t\wedge \tau_n}^{\tau_n} g(s, u_s) d\langle M\rangle_s\right)\bigg|\mathcal{F}_t\right].$$

So $Y_t^n = \text{ess sup}_{u \in U} H_t^n(u)$ and $Y_t^{n+1} = \text{ess sup}_{u \in U} H_t^{n+1}(u)$. We want to show that $Y_t^n \leq Y_t^{n+1}$ and for this it is sufficient to show that for any $\hat{u} \in U$, $H_t^n(\hat{u}) \leq Y_t^{n+1}$;

$$H_t^n(\hat{u}) = E\left[\mathcal{E}_{t \wedge \tau_n, \tau_n}\left(\int f_l'(\hat{u})dM\right)\left(\eta^n + \int_{t \wedge \tau_n}^{\tau_n} g(s, \hat{u}_s)d\langle M\rangle_s\right)\Big|\mathcal{F}_t\right]$$

$$= E\left[\mathcal{E}_{t \wedge \tau_n, \tau_n}\left(\int f_l'(\hat{u})dM\right)E[\eta^n|\mathcal{F}_{t \vee \tau_n}]\Big|\mathcal{F}_t\right]$$

$$+ E\left[\mathcal{E}_{t \wedge \tau_n, \tau_n}\left(\int f_l'(\hat{u})dM\right)\int_{t \wedge \tau_n}^{\tau_n} g(s, \hat{u}_s)d\langle M\rangle_s\Big|\mathcal{F}_t\right]$$

$$\leq E\left[\mathcal{E}_{t \wedge \tau_n, \tau_n}\left(\int f_l'(\hat{u})dM\right)E[\eta^{n+1}|\mathcal{F}_{t \vee \tau_n}]\Big|\mathcal{F}_t\right]$$

$$+ E\left[\mathcal{E}_{t \wedge \tau_n, \tau_n}\left(\int f_l'(\hat{u})dM\right)\int_{t \wedge \tau_n}^{\tau_n} g(s, \hat{u}_s)d\langle M\rangle_s\Big|\mathcal{F}_t\right]$$

and for Y_t^{n+1} we have the representation

$$Y_t^{n+1} = \text{ess sup}_{u \in U} E\left[\mathcal{E}_{t \wedge \tau_{n+1}, \tau_{n+1}}\left(\int f_l'(u)dM\right)\left(\eta^{n+1} + \int_{t \wedge \tau_{n+1}}^{\tau_{n+1}} g(s, u_s)d\langle M\rangle_s\right)\Big|\mathcal{F}_t\right] =$$

$$\text{ess sup}_{u \in U}\left[\mathcal{E}_{t \wedge \tau_{n+1}, \tau_{n+1}}\left(\int f_l'(u)dM\right)\left(\eta^{n+1} + \int_{t \wedge \tau_n}^{\tau_n} g(s, u_s)d\langle M\rangle_s + \int_{(t \vee \tau_n) \wedge \tau_{n+1}}^{\tau_{n+1}} g(s, u_s)d\langle M\rangle_s\right)\Big|\mathcal{F}_t\right] =$$

$$= \text{ess sup}_{u \in U}\left\{E\left[\mathcal{E}_{t \wedge \tau_{n+1}, \tau_{n+1}}\left(\int f_l'(u)dM\right)\int_{t \wedge \tau_n}^{\tau_n} g(s, u_s)d\langle M\rangle_s\Big|\mathcal{F}_t\right]+\right.$$

$$\left.+ E\left[\mathcal{E}_{t \wedge \tau_{n+1}, \tau_{n+1}}\left(\int f_l'(u)dM\right)\left(\eta^{n+1} + \int_{(t \vee \tau_n) \wedge \tau_{n+1}}^{\tau_{n+1}} g(s, u_s)d\langle M\rangle_s\right)\Big|\mathcal{F}_t\right]\right\} =$$

$$= \text{ess sup}_{u \in U}\left\{E\left[\mathcal{E}_{t \wedge \tau_n, \tau_n}\left(\int f_l'(u)dM\right)\int_{t \wedge \tau_n}^{\tau_n} g(s, u_s)d\langle M\rangle_s\Big|\mathcal{F}_t\right]+\right.$$

$$+ E\left[\mathcal{E}_{t \wedge \tau_n, \tau_n}\left(\int f_l'(u)dM\right)E\left[\mathcal{E}_{(t \vee \tau_n) \wedge \tau_{n+1}, \tau_{n+1}}\left(\int f_l'(u)dM\right)\times\right.\right.$$

$$\left.\left.\times\left(\eta^{n+1} + \int_{(t \vee \tau_n) \wedge \tau_{n+1}}^{\tau_{n+1}} g(s, u_s)d\langle M\rangle_s\right)\Big|\mathcal{F}_{t \vee \tau_n}\right]\Big|\mathcal{F}_t\right]\right\} =$$

$$= \text{ess sup}_{u \in U}\left\{E\left[\mathcal{E}_{t \wedge \tau_n, \tau_n}\left(\int f_l'(u)dM\right)H_{t \vee \tau_n}^{n+1}(u)\Big|\mathcal{F}_t\right]+\right.$$

$$\left.+ E\left[\mathcal{E}_{t \wedge \tau_n, \tau_n}\left(\int f_l'(u)dM\right)\int_{t \wedge \tau_n}^{\tau_n} g(s, u_s)d\langle M\rangle_s\Big|\mathcal{F}_t\right]\right\}.$$

Define the class of controls $U_{\tau_n}^T$: $U_{\tau_n}^T = \{u \in U : u_t 1_{(t \leq \tau_n)} = \hat{u}_t 1_{(t \leq \tau_n)}\}$. Now because $U_{\tau_n}^T \subset U$ and $H_t^n(u)$ has the lattice property ([17]) we obtain the following chain of inequalities

$$Y_t^{n+1} \geq \text{ess sup}_{u \in U_{\tau_n}^T}\left\{E\left[\mathcal{E}_{t \wedge \tau_n, \tau_n}\left(\int f_l'(u)dM\right)H_{t \vee \tau_n}^{n+1}(u)\Big|\mathcal{F}_t\right]\right.$$

$$\left.+ E\left[\mathcal{E}_{t \wedge \tau_n, \tau_n}\left(\int f_l'(u)dM\right)\int_{t \wedge \tau_n}^{\tau_n} g(s, u_s)d\langle M\rangle_s\Big|\mathcal{F}_t\right]\right\}$$

$$
\begin{aligned}
&= \operatorname{ess\,sup}_{u \in U_{\tau_n}^T} E\left[\mathcal{E}_{t \wedge \tau_n, \tau_n}\left(\int f_l'(u)dM\right) H_{t \vee \tau_n}^{n+1}(u)\Big|\mathcal{F}_t\right] \\
&\quad + E\left[\mathcal{E}_{t \wedge \tau_n, \tau_n}\left(\int f_l'(\hat{u})dM\right) \int_{t \wedge \tau_n}^{\tau_n} g(s, \hat{u}_s)d\langle M\rangle_s\Big|\mathcal{F}_t\right] \\
&= E\left[\mathcal{E}_{t \wedge \tau_n, \tau_n}\left(\int f_l'(\hat{u})dM\right) \operatorname{ess\,sup}_{u \in U_{\tau_n}^T} H_{t \vee \tau_n}^{n+1}(u)\Big|\mathcal{F}_t\right] \\
&\quad + E\left[\mathcal{E}_{t \wedge \tau_n, \tau_n}\left(\int f_l'(\hat{u})dM\right) \int_{t \wedge \tau_n}^{\tau_n} g(s, \hat{u}_s)d\langle M\rangle_s\Big|\mathcal{F}_t\right] \\
&= E\left[\mathcal{E}_{t \wedge \tau_n, \tau_n}\left(\int f_l'(\hat{u})dM\right) Y_{t \vee \tau_n}^{n+1}\Big|\mathcal{F}_t\right] \\
&\quad + E\left[\mathcal{E}_{t \wedge \tau_n, \tau_n}\left(\int f_l'(\hat{u})dM\right) \int_{t \wedge \tau_n}^{\tau_n} g(s, \hat{u}_s)d\langle M\rangle_s\Big|\mathcal{F}_t\right] \\
&\geq E\left[\mathcal{E}_{t \wedge \tau_n, \tau_n}\left(\int f_l'(\hat{u})dM\right) E[\eta^{n+1}|\mathcal{F}_{t \vee \tau_n}]\Big|\mathcal{F}_t\right] \\
&\quad + E\left[\mathcal{E}_{t \wedge \tau_n, \tau_n}\left(\int f_l'(\hat{u})dM\right) \int_{t \wedge \tau_n}^{\tau_n} g(s, \hat{u}_s)d\langle M\rangle_s\Big|\mathcal{F}_t\right] \geq H_t^n(\hat{u}).
\end{aligned}
$$

This implies that $Y_t^n \leq Y_t^{n+1}$. $\square$

Now define the process $Y_t := \sup_n Y_t^n = \lim_n Y_t^n$. Our aim is to show that $V_t = Y_t$, but first we prove that $V_t \leq Y_t$.

For this we need the following

Lemma 3.5. *For any $u \in U$ and $t \in [0; T]$, the family of random variables*

$$\left\{\xi_n = \mathcal{E}_{t \wedge \tau_n, \tau_n}\left(\int f_l'(u)dM\right)\left(\eta^n + \int_{t \wedge \tau_n}^{\tau_n} g(s, u_s)d\langle M\rangle_s\right)\right\}_{n \geq 1}$$

is uniformly integrable.

Proof. We know that for any $u \in U$, $\int f_l'(u)dM \in BMO$ and

$$\mathcal{E}_{t \wedge \tau_n, \tau_n}\left(\int f_l'(u)dM\right) = \mathcal{E}_{t,T}\left(\int f_l'(u)dM^n\right)$$

where $M^n = M^{\tau_n}$. Because $\int_0^t f_l'(s, u_s)dM_s^n = \int_0^{t \wedge \tau_n} f_l'(s, u_s)dM_s$ for any $n \in N$ the following inequality holds

$$\left\|\int f_l'(u)dM^n\right\|_{BMO_2} \leq \left\|\int f_l'(u)dM\right\|_{BMO_2} < \infty. \tag{36}$$

Using Kazamaki's [21] result we obtain that there exist $p > 1$ and $C_p > 0$ such that for any $n \in N$,

$$E\left[\mathcal{E}_{t,T}^p\left(\int f_l'(u)dM^n\right)\Big|\mathcal{F}_t\right] \leq C_p,$$

where p and C_p are independent of n.

Now let us take $0 < \epsilon < p - 1$, $\tilde{p} = \frac{p}{1+\epsilon} > 1$ and $\tilde{q}$ such that $\frac{1}{\tilde{p}} + \frac{1}{\tilde{q}} = 1$.

Using the Hölder inequality we obtain

$$\sup_n E|\xi_n|^{1+\epsilon} \leq \sup_n \left\{ E\left|\mathcal{E}_{t,T}\left(\int f'_l(u)dM^n\right)\right|^{1+\epsilon} \cdot \left|\eta^n + \int_{t\wedge\tau_n}^{\tau_n} g(s,u_s)d\langle M\rangle_s\right|^{1+\epsilon}\right\}$$

$$\leq \sup_n \left[\left\{E\left|\mathcal{E}_{t,T}\left(\int f'_l(u)dM^n\right)\right|^{(1+\epsilon)\tilde{p}}\right\}^{\frac{1}{\tilde{p}}} \cdot \left\{E\left|\eta^n + \int_{t\wedge\tau_n}^{\tau_n} g(s,u_s)d\langle M\rangle_s\right|^{(1+\epsilon)\tilde{q}}\right\}^{\frac{1}{\tilde{q}}}\right]$$

$$= \sup_n \left[\left\{E\left[\mathcal{E}^p_{t,T}\left(\int f'_l(u)dM^n\right)\right]\right\}^{\frac{1}{\tilde{p}}} \cdot \left\{E\left|\eta^n + \int_{t\wedge\tau_n}^{\tau_n} g(s,u_s)d\langle M\rangle_s\right|^{(1+\epsilon)\tilde{q}}\right\}^{\frac{1}{\tilde{q}}}\right]$$

$$\leq (C_p)^{\frac{1}{\tilde{p}}} \cdot \sup_n \left\{2^{(1+\epsilon)\tilde{q}-1}\left[E|\eta^n|^{(1+\epsilon)\tilde{q}} + E\left(\int_{t\wedge\tau_n}^{\tau_n} |g(s,u_s)|d\langle M\rangle_s\right)^{(1+\epsilon)\tilde{q}}\right]\right\}^{\frac{1}{\tilde{q}}}$$

$$\leq 2^{1+\epsilon-\frac{1}{\tilde{q}}} \cdot (C_p)^{\frac{1}{\tilde{p}}} \cdot \left\{E|\eta|^{(1+\epsilon)\tilde{q}} + E\left(\int_0^T |g(s,u_s)|d\langle M\rangle_s\right)^{(1+\epsilon)\tilde{q}}\right\}^{\frac{1}{\tilde{q}}}.$$

According to condition **3)** of Theorem 3.1, η is integrable to an exponential degree, so $E|\eta|^{(1+\epsilon)\tilde{q}} < \infty$. With this we need $E\left(\int_0^T |g(s,u_s)|d\langle M\rangle_s\right)^{(1+\epsilon)\tilde{q}}$ to be finite. Because $u \in U$, there exists a constant $D \geq 0$ such that $|u_s| \leq D$. Therefore

$$E\left(\int_0^T |g(s,u_s)|d\langle M\rangle_s\right)^{(1+\epsilon)\tilde{q}} = E\left(\int_0^T |f(s,u_s) - f'_l(s,u_s)u_s|d\langle M\rangle_s\right)^{(1+\epsilon)\tilde{q}}$$

$$\leq 2^{(1+\epsilon)\tilde{q}-1}\left[E\left(\int_0^T |f(s,u_s)|d\langle M\rangle_s\right)^{(1+\epsilon)\tilde{q}} + D^{(1+\epsilon)\tilde{q}}E\left(\int_0^T |f'_l(s,u_s)|d\langle M\rangle_s\right)^{(1+\epsilon)\tilde{q}}\right].$$

Now, using the quadratic growth condition for the generator f, Lemma 3.13 (see Appendix) and the fact that the martingales M and $\int \alpha dM$ are from the class BMO we obtain from the **energy inequality** ([21]):

$$E\left(\int_0^T |f(s,u_s)|d\langle M\rangle_s\right)^{(1+\epsilon)\tilde{q}} \leq E\left(\int_0^T \left(\alpha_s + \frac{\gamma}{2}D^2\right)d\langle M\rangle_s\right)^{(1+\epsilon)\tilde{q}} =$$

$$= E\left(\int_0^T \alpha_s d\langle M\rangle_s + \frac{\gamma}{2}D^2\langle M\rangle_T\right)^{(1+\epsilon)\tilde{q}} < \infty$$

and

$$E\left(\int_0^T |f'_l(s,u_s)|d\langle M\rangle_s\right)^{(1+\epsilon)\tilde{q}} \leq E\left(2\int_0^T \alpha_s d\langle M\rangle_s + \left(\gamma + \frac{3\gamma D^2}{2}\right)\langle M\rangle_T\right)^{(1+\epsilon)\tilde{q}} < \infty.$$

Finally, we obtain that $\sup_n E|\xi_n|^{1+\epsilon} < \infty$ for some $\varepsilon > 0$, which means that the family $\{\xi_n\}_{n\geq 1}$ is uniformly integrable. □

Lemma 3.6. $V_t \leq Y_t$ P a.s.

Proof. Using Lemma 3.5 we obtain

$$Y_t = \lim_n Y_t^n = \lim_n \operatorname{ess\,sup}_{u \in U} E\left[\mathcal{E}_{t \wedge \tau_n, \tau_n}\left(\int f_l'(u)dM\right)\left(\eta^n + \int_{t \wedge \tau_n}^{\tau_n} g(s, u_s)d\langle M\rangle_s\right)\Big|\mathcal{F}_t\right] \geq$$

$$\geq \operatorname{ess\,sup}_{u \in U} \lim_n E\left[\mathcal{E}_{t \wedge \tau_n, \tau_n}\left(\int f_l'(u)dM^n\right)\left(\eta^n + \int_{t \wedge \tau_n}^{\tau_n} g(s, u_s)d\langle M\rangle_s\right)\Big|\mathcal{F}_t\right] =$$

$$= \operatorname{ess\,sup}_{u \in U} E\left[\mathcal{E}_{t,T}\left(\int f_l'(u)dM\right)\left(\eta + \int_t^T g(s, u_s)d\langle M\rangle_s\right)\Big|\mathcal{F}_t\right] = V_t. \quad \square$$

To prove that $Y_t = V_t$, we only have to show the reverse inequality: $Y_t \leq V_t$. For this we need to show that V is a supermartingale.

Lemma 3.7. *The process V is a supermartingale.*

Proof. According to Section 3.1, V is a supersolution of equation (1) and because the generator f is nonnegative it follows that V is a local supermartingale. This means that there exist stopping times $(\varsigma_m)_{m \geq 1}$ such that for any m and $t > s$ we have the inequality: $E[V_{t \wedge \varsigma_m}|\mathcal{F}_s] \leq V_{s \wedge \varsigma_m}$. To prove the inequality $E[V_t|\mathcal{F}_s] \leq V_s$ we need Fatou's Lemma and $\lim_m V_{\varsigma_m} = \eta$. We show this using the following Lemma.

Lemma 3.8. *The a priori estimate for the value process V is*

$$E\left[\mathcal{E}_{t,T}\left(\int f_l'(0)dM\right)\left(\eta + \int_t^T f(s,0)d\langle M\rangle_s\right)\Big|\mathcal{F}_t\right] \leq V_t \leq \frac{1}{\gamma} \ln E\left[e^{\gamma\eta + \gamma \int_t^T \alpha_s d\langle M\rangle_s}\Big|\mathcal{F}_t\right].$$

Proof. According to Lemma 3.3, for any $n \in N$ the following inequality holds:

$$Y_t^n \leq \frac{1}{\gamma} \ln E\left[e^{\gamma\eta^n + \gamma \int_t^T 1_{(s \leq \tau_n)} \alpha_s d\langle M\rangle_s}\Big|\mathcal{F}_t\right] = \frac{1}{\gamma} \ln E\left[e^{\gamma\eta^n + \gamma \int_{t \wedge \tau_n}^{\tau_n} \alpha_s d\langle M\rangle_s}\Big|\mathcal{F}_t\right].$$

Now, by condition **3)** of Theorem 3.1 we can take the limits on both sides of the inequality as $n \to \infty$. As a result we obtain

$$Y_t \leq \frac{1}{\gamma} \ln E\left[e^{\gamma\eta + \gamma \int_t^T \alpha_s d\langle M\rangle_s}\Big|\mathcal{F}_t\right].$$

Because $V_t \leq Y_t$ the last inequality is true for the value process too.

Finally, if in the representation of V we insert the control $u^0 \equiv 0$, then we obtain the inequality

$$V_t \geq E\left[\mathcal{E}_{t,T}\left(\int f_l'(0)dM\right)\left(\eta + \int_t^T f(s,0)d\langle M\rangle_s\right)\Big|\mathcal{F}_t\right]. \quad \square$$

Now, if take the limit as $t \to T$ in the a priori estimate of V we obtain $\eta \leq \lim_{t \to T} V_t \leq \eta$. This means that the value process V is continuous in T and $\eta = \lim_m V_{\varsigma_m}$.

Now, because f and η are nonnegative it follows from Lemma 3.8 that V is nonnegative too. Using Fatou's Lemma we obtain

$$E[V_t|\mathcal{F}_s] = E[\liminf_m V_{t \wedge \varsigma_m}|\mathcal{F}_s] \leq \liminf_m E[V_{t \wedge \varsigma_m}|\mathcal{F}_s] \leq \liminf_m V_{s \wedge \varsigma_m} = V_s.$$

This means that V is a supermartingale. $\quad \square$

Now we are ready to prove that $V_t = Y_t$ P a.s.

Lemma 3.9. $V_t = Y_t$ P a.s.

Proof. To prove Lemma 3.9 it is sufficient to show that for any n, $Y^n \leq V$. We can show this in exactly the same way as it was done in Lemma 3.4 taking $\tau_{n+1} = T$ and η instead of η^{n+1}. □

Now it is time to prove that Y is a solution of equation (1). According to Lemma 3.3 for any $n \in N$ the following inequality holds:

$$Y_t^n \leq \frac{1}{\gamma} \ln E\left[e^{\gamma\eta^n + \gamma \int_t^T 1_{(s \leq \tau_n)} \alpha_s d\langle M\rangle_s} \Big| \mathcal{F}_t\right] \leq \frac{1}{\gamma} \ln E\left[e^{\gamma\eta^n + \gamma \int_0^{\tau_n} \alpha_s d\langle M\rangle_s} \Big| \mathcal{F}_t\right].$$

By condition **3)** of Theorem 3.1 we can take the limits on both sides of the inequality when $n \to \infty$. As a result we obtain:

$$Y_t \leq \frac{1}{\gamma} \ln E\left[e^{\gamma\eta + \gamma \int_0^T \alpha_s d\langle M\rangle_s} \Big| \mathcal{F}_t\right].$$

On the other hand, because Y^n is a supermartingale, $Y_t^n \geq E[Y_T^n | \mathcal{F}_t] = E[\eta^n | \mathcal{F}_t] \geq 0$.

So we have an estimate

$$0 \leq Y_t^n \leq Y_t \leq \frac{1}{\gamma} \ln E\left[e^{\gamma\eta + \gamma \int_0^T \alpha_s d\langle M\rangle_s} \Big| \mathcal{F}_t\right].$$

Define stopping times: $\sigma_k = \inf\left\{t \geq 0 : \frac{1}{\gamma} \ln E\left[e^{\gamma\eta + \gamma \int_0^T \alpha_s d\langle M\rangle_s} \Big| \mathcal{F}_t\right] \geq k\right\}$. It is obvious that $|Y_{t \wedge \sigma_k}| \leq k$ and $|Y_{t \wedge \sigma_k}^n| \leq k$. From (35) we know that

$$Y_t^n = Y_0^n - \int_0^t 1_{(s \leq \tau_n)} f(s, \varphi_s^n) d\langle M\rangle_s + \int_0^t \varphi_s^n dM_s + L_t^n.$$

From this we have

$$Y_{t \wedge \sigma_k}^n = Y_0^n - \int_0^t 1_{(s \leq \sigma_k \wedge \tau_n)} f(s, 1_{(s \leq \sigma_k)} \varphi_s^n) d\langle M\rangle_s + \int_0^t 1_{(s \leq \sigma_k)} \varphi_s^n dM_s + L_{t \wedge \sigma_k}^n.$$

Now define the following processes:

$$Y_k^n(t) = Y_{t \wedge \sigma_k}^n, \qquad \varphi_k^n(t) = 1_{(t \leq \sigma_k)} \varphi_t^n, \qquad L_k^n(t) = L_{t \wedge \sigma_k}^n.$$

We obtain that the triple $(Y_k^n; \varphi_k^n; L_k^n)$ satisfies the equation

$$\begin{cases} Y_k^n(t) = Y_k^n(0) - \int_0^t 1_{(s \leq \sigma_k \wedge \tau_n)} f(s, \varphi_k^n(s)) d\langle M\rangle_s + \int_0^t \varphi_k^n(s) dM_s + L_k^n(t), \\ Y_k^n(T) = Y_{\sigma_k}^n. \end{cases} \quad (37)$$

Because f is nonnegative, $1_{(s \leq \sigma_k \wedge \tau_n)} f(s, x) \uparrow 1_{(s \leq \sigma_k)} f(s, x)$ and $Y_{\sigma_k}^n \uparrow Y_{\sigma_k}$. According to Lemma 3.4, $Y_k^n \leq Y_k^{n+1}$, so by the Lemma 3.2 (Monotonne Stability Lemma), there exists a triple $(Y_k; \varphi_k; L_k)$ which satisfies the equation

$$\begin{cases} Y_k(t) = Y_k(0) - \int_0^t 1_{(s \leq \sigma_k)} f(s, \varphi_k(s)) d\langle M\rangle_s + \int_0^t \varphi_k(s) dM_s + L_k(t), \\ Y_k(T) = Y_{\sigma_k}. \end{cases} \quad (38)$$

It is clear that $Y_k(t) = \sup_n Y_k^n(t) = \sup_n Y_{t\wedge\sigma_k}^n = Y_{t\wedge\sigma_k}$ and

$$1_{(s\leq\sigma_k)}\varphi_{k+1}(s) = 1_{(s\leq\sigma_k)}\lim_n \varphi_{k+1}^n(s) = 1_{(s\leq\sigma_k)}\lim_n 1_{(s\leq\sigma_{k+1})}\varphi_s^n$$
$$= 1_{(s\leq\sigma_k)}\lim_n 1_{(s\leq\sigma_k)}\varphi_s^n = 1_{(s\leq\sigma_k)}\lim_n \varphi_k^n(s) = 1_{(s\leq\sigma_k)}\varphi_k(s).$$

So we can correctly define the process $\varphi_t := \varphi_k(t)$ when $t \leq \sigma_k$.

Now we can rewrite equation (38) as

$$Y_{t\wedge\sigma_k} = Y_0 - \int_0^{t\wedge\sigma_k} f(s, \varphi_s)d\langle M\rangle_s + \int_0^{t\wedge\sigma_k} \varphi_s dM_s + L_k(t).$$

If we take the limit as $k \to \infty$, then we obtain that on the stochastic interval $[[0; T[[$

$$Y_t = Y_0 - \int_0^t f(s, \varphi_s)d\langle M\rangle_s + \int_0^t \varphi_s dM_s + L_t. \tag{39}$$

So it remains only to prove that $\eta = Y_T = \lim_{t \to T} Y_t$.

But this is obvious because we know that $Y_t = V_t$ and V_t is continuous in T. $\square$

Remark 3.1. The proof of Proposition 3.5 is exactly the same if we take η bounded from below by a constant.

3.4 The general case: the proof of the Theorem 3.1

In this Section we sketch the proof of Theorem 3.1 because it is similar to the proof in Section 3.3.

Now consider equation (1) and assume that the conditions of Theorem 3.1 hold. Consider the new measure $d\tilde{P} = \mathcal{E}_T\left(\int f'_l(0)dM\right)dP$ and the generator $h(s,x) = f(s,x) - f'_l(s,0)x - f(s,0)$. Because f is convex, h is a nonnegative and convex generator. According to Girsanov's Theorem and Kazamaki's [21] result, the process $\tilde{M}_t = M_t - \int_0^t f'_l(s,0)d\langle M\rangle_s$ is a $BMO(\tilde{P})$ martingale. Now with equation (1) let us consider equation ($\tilde{1}$):

$$\begin{cases} Y_t = Y_0 - \int_0^t h(s, Z_s)d\langle \tilde{M}\rangle_s + \int_0^t Z_s d\tilde{M}_s + L_t, \\ Y_T = \eta + \int_0^T f(s,0)d\langle M\rangle_s. \end{cases} \tag{$\tilde{1}$}$$

Lemma 3.10. a) *The triple* (Y_t, φ_t, L_t) *is a solution of equation (1) if and only if the triple* $(\tilde{Y}_t = Y_t + \int_0^t f(s,0)d\langle M\rangle_s, \varphi_t, L_t)$ *is a solution of equation ($\tilde{1}$).*

b) *If* $\|\eta\|_\infty + \|\int_0^T \alpha_s d\langle M\rangle_s\|_\infty < \infty$, *then for any bounded solution of equation ($\tilde{1}$) the following inequality holds:*

$$\tilde{Y}_t \leq \frac{1}{\gamma} \ln E\left[e^{\gamma\eta + \gamma\int_t^T \alpha_s d\langle M\rangle_s}\bigg|\mathcal{F}_t\right] + \int_0^t \alpha_s d\langle M\rangle_s. \tag{40}$$

Proof. The proof of the part **a)** follows from Girsanov's theorem.

For the part **b)**: if $\tilde{Y}_t$ is a bounded solution of equation ($\tilde{1}$), then according to the part **a)**, $\tilde{Y}_t - \int_0^t f(s,0)d\langle M\rangle_s$ is the bounded solution of equation (1). Now using Lemma 3.3 we obtain the following estimation for $\tilde{Y}_t - \int_0^t f(s,0)d\langle M\rangle_s$:

$$\tilde{Y}_t - \int_0^t f(s,0)d\langle M\rangle_s \leq \frac{1}{\gamma} \ln E\left[e^{\gamma\eta + \gamma\int_t^T \alpha_s d\langle M\rangle_s}\bigg|\mathcal{F}_t\right]$$

which gives inequality (40). $\square$

The corresponding value process for equation (1̃) has the form:
$$\tilde{V}_t = \operatorname{ess\,sup}_{u \in U} \tilde{E}\left[\mathcal{E}_{t,T}\left(\int h'_l(u)d\tilde{M}\right)\left(\eta + \int_0^T f(s,0)d\langle M\rangle_s + \int_t^T [h(s,u_s) - h'_l(s,u_s)u_s]d\langle \tilde{M}\rangle_s\right)\bigg|\mathcal{F}_t\right].$$

The next lemma shows the connection between the value processes for equations (1) and (1̃):

Lemma 3.11. $\tilde{V}_t - \int_0^t f(s,0)d\langle M\rangle_s = V_t$,

where V_t is the corresponding value process for equation (1).

Proof. First of all notice that $h'_l(s,u_s) = f'_l(s,u_s) - f'_l(s,0)$ and

$$h(s,u_s) - h'_l(s,u_s)u_s = f(s,u_s) - f'_l(s,0)u_s - f(s,0) - f'_l(s,u_s)u_s + f'_l(s,0)u_s =$$
$$= f(s,u_s) - f'_l(s,u_s)u_s - f(s,0).$$

After that, we need to simplify the expression $\mathcal{E}_{t,T}\left(\int h'_l(u)d\tilde{M}\right)$:

$$\mathcal{E}_{t,T}\left(\int h'_l(u)d\tilde{M}\right) = \exp\left\{\int_t^T h'_l(s,u_s)d\tilde{M}_s - \frac{1}{2}\int_t^T |h'_l(s,u_s)|^2 d\langle M\rangle_s\right\} =$$

$$= \exp\left\{\int_t^T f'_l(s,u_s)d\tilde{M}_s - \int_t^T f'_l(s,0)d\tilde{M}_s - \frac{1}{2}\int_t^T |f'_l(s,u_s) - f'_l(s,0)|^2 d\langle M\rangle_s\right\} =$$

$$= \exp\left\{\int_t^T f'_l(s,u_s)dM_s - \int_t^T f'_l(s,u_s)f'_l(s,0)d\langle M\rangle_s - \int_t^T f'_l(s,0)dM_s + \int_t^T |f'_l(s,0)|^2 d\langle M\rangle_s -\right.$$
$$\left.-\frac{1}{2}\int_t^T |f'_l(s,u_s)|^2 d\langle M\rangle_s + \int_t^T f'_l(s,u_s)f'_l(s,0)d\langle M\rangle_s - \frac{1}{2}\int_t^T |f'_l(s,0)|^2 d\langle M\rangle_s\right\} =$$

$$= \exp\left\{\int_t^T f'_l(s,u_s)dM_s - \frac{1}{2}\int_t^T |f'_l(s,u_s)|^2 d\langle M\rangle_s - \left(\int_t^T f'_l(s,0)dM_s - \frac{1}{2}\int_t^T |f'_l(s,0)|^2 d\langle M\rangle_s\right)\right\} =$$

$$= \frac{\mathcal{E}_{t,T}\left(\int f'_l(u)dM\right)}{\mathcal{E}_{t,T}\left(\int f'_l(0)dM\right)}.$$

Using this expression we obtain

$$\tilde{V}_t - \int_0^t f(s,0)d\langle M\rangle_s = \operatorname{ess\,sup}_{u \in U} \tilde{E}\left[\mathcal{E}_{t,T}\left(\int h'_l(u)d\tilde{M}\right)\left(\eta + \int_0^T f(s,0)d\langle M\rangle_s\right.\right.$$
$$\left.\left.+ \int_t^T [h(s,u_s) - h'_l(s,u_s)u_s]d\langle M\rangle_s\right)\bigg|\mathcal{F}_t\right] - \int_0^t f(s,0)d\langle M\rangle_s$$

$$= \operatorname{ess\,sup}_{u \in U} \tilde{E}\left[\frac{\mathcal{E}_{t,T}\left(\int f'_l(u)dM\right)}{\mathcal{E}_{t,T}\left(\int f'_l(0)dM\right)}\left(\eta + \int_0^T f(s,0)d\langle M\rangle_s\right.\right.$$
$$\left.\left.+ \int_t^T [f(s,u_s) - f'_l(s,u_s)u_s - f(s,0)]d\langle M\rangle_s\right)\bigg|\mathcal{F}_t\right] - \int_0^t f(s,0)d\langle M\rangle_s$$

$$= \operatorname{ess\,sup}_{u \in U} E\left[\mathcal{E}_{t,T}\left(\int f'_l(u)dM\right)\left(\eta + \int_0^t f(s,0)d\langle M\rangle_s\right.\right.$$

$$+ \int_t^T [f(s, u_s) - f'_l(s, u_s)u_s]d\langle M\rangle_s\bigg)\bigg|\mathcal{F}_t\bigg] - \int_0^t f(s,0)d\langle M\rangle_s$$

$$= \operatorname{ess\,sup}_{u \in U} E^u\bigg[\eta + \int_t^T [f(s, u_s) - f'_l(s, u_s)u_s]d\langle M\rangle_s\bigg|\mathcal{F}_t\bigg] = V_t. \quad \Box$$

So, according to Lemma 3.10 and Lemma 3.11, we need only to show that the value process $\tilde{V}_t$ is a solution of equation $(\tilde{1})$.

This can be shown in the same way as in Section 3.3 for nonnegative η and f using a priori estimate (40) for any bounded solution of equation $(\tilde{1})$. **Theorem 3.1** is proved. $\quad \Box$

3.5 The multidimensional case for equation (1)

Let M be a d-dimensional square integrable martingale defined on a filtered probability space $(\Omega, \mathcal{F}, (\mathcal{F}_t)_{0 \le t \le T}, P)$ satisfying the usual conditions. Note that the quadratic variation $\langle M\rangle$ is a matrix with components $\langle M^i, M^j\rangle$ $i, j \in \{1, ..., d\}$. Recalling that each component $d\langle M^i, M^j\rangle$ is absolutely continuous with respect to $dC = \sum_i d\langle M^i\rangle$ and there exists a predictable process m taking values in $R^{d\times d}$ such that $d\langle M\rangle$ can be written in the following form: $d\langle M\rangle_s = m_s m'_s dC_s$.

We consider a backward stochastic differential equation of the form:

$$\begin{cases} Y_t = Y_0 - \int_0^t f(s, Z_s)dC_s + \int_0^t Z_s dM_s + L_t, \\ Y_T = \eta. \end{cases} \quad (41)$$

where the generator $f : [0; T] \times \Omega \times R^d \to R$ is a measurable function; $f(\cdot, \cdot, z)$ is a predictable process for any $z = (z^1, ..., z^d)$ and η is an $\mathcal{F}_T$-measurable random variable. The couple (f, η) is called parameters of equation (41).

We say that a continuous R^d valued martingale M is from the class BMO if for any $i = 1, ..., d$ $\sup_\tau \big\|E[\langle M^i\rangle_T - \langle M^i\rangle_\tau|\mathcal{F}_\tau]\big\|_\infty < \infty$ where the sup is taken over all stopping times $0 \le \tau \le T$.

Now we can formulate the main theorem

Theorem 3.2. Suppose that the filtration $\{\mathcal{F}_t\}_{0 \le t \le T}$ is continuous and $\{M_t\}_{0 \le t \le T}$ is a martingale from the class BMO. Let the parameters (f, η) of equation (41) satisfy the following conditions:
1) for any (t, ω), $f(t, \omega, \cdot)$ is a continuous and convex function.
2) there exist a predictable non-negative process α_t and a constant $\gamma \ge 0$ such that $\sup_\tau \big\|E[\int_\tau^T \alpha_s dC_s|\mathcal{F}_\tau]\big\|_\infty < \infty$ and for any (t, ω, z)

$$|f(t, \omega, z)| \le \alpha_t(\omega) + \frac{\gamma}{2}|z|^2.$$

3) $Ee^{\gamma\eta + \gamma \int_0^T \alpha_s dC_s} < \infty$ and $\eta + \int_0^T f(s,0)dC_s \ge -D$ for some $D \ge 0$.

Then there exists a solution $V = \{V_t\}_{0 \le t \le T}$ of equation (41) represented in the form

$$V_t = \operatorname{ess\,sup}_{u \in U} E\bigg[\mathcal{E}_{t,T}\bigg(\int f'(u)dM\bigg)\bigg(\eta + \int_t^T [f(s, u_s) - f'(s, u_s)u_s]dC_s\bigg)\bigg|\mathcal{F}_t\bigg]$$

where f' is a measureble version of subdifferential of the function f and U is the class of predictable, bounded R^d valued controls:

$$U = \{u : \exists C_u \ge 0, |u_t| \le C_u\}.$$

The proof of **Theorem 3.2** is the same as the proof of **Theorem 3.1**.

3.6 Uniqueness of the solution

In this section our aim is to prove that in a special class of processes there exists the unique solution of equation (1).
Let us define the class of solutions:

$$\aleph = \left\{ (Y, Z, L) \ : \ Ee^{p(Y^+ + \int \alpha d\langle M\rangle)^*} < \infty \ \text{ and } \ Ee^{\epsilon(Y^-)^*} < \infty \right\}$$

for some $p > \gamma$ and $\epsilon > 0$, where γ is a constant from condition **2)** of **Theorem 3.1**.
As we know from the previous section, there exists the triple (V, φ, N) satisfying equation (1) where the first component V coincides with the value process:

$$V_t = \operatorname*{ess\,sup}_{u \in U} E^u \left[\eta + \int_t^T [f(s, u_s) - f'(s, u_s) u_s] d\langle M\rangle_s \bigg| \mathcal{F}_t \right],$$

where U is the class of predictable, bounded controls.

Proposition 3.6. *If* $Ee^{p\eta^+ + p \int_0^T \alpha_s d\langle M\rangle_s} < \infty$ *then* V *is from the class* $\aleph$.

Proof. According to the **subsection 3.2** and Lemma 3.3

$$\operatorname*{ess\,sup}_{u \in U} E\left[\mathcal{E}_{t \wedge \tau_n, \tau_n}\left(\int f'_l(u) dM\right)\left(\eta^n + \int_{t \wedge \tau_n}^{\tau_n} g(s, u_s) d\langle M\rangle_s\right) \bigg| \mathcal{F}_t\right] \leq$$

$$\leq \frac{1}{\gamma} \ln E\left[e^{\gamma \eta^n + \gamma \int_{t \wedge \tau_n}^{\tau_n} \alpha_s d\langle M\rangle_s} \bigg| \mathcal{F}_t\right], \tag{42}$$

where $\eta^n = \eta^+ \wedge n - \eta^- \wedge n$ and $\tau_n = \inf\left\{t \geq 0 \ : \ \int_0^t \alpha_s^2 d\langle M\rangle_s \vee \langle M\rangle_t \geq n\right\}$. Because

$$e^{\gamma \eta^n + \gamma \int_{t \wedge \tau_n}^{\tau_n} \alpha_s d\langle M\rangle_s} \leq e^{\gamma \eta^+ + \gamma \int_0^T \alpha_s d\langle M\rangle_s}$$

and $Ee^{\gamma \eta^+ + \gamma \int_0^T \alpha_s d\langle M\rangle_s} < \infty$ applying Fatou's Lemma we obtain

$$\limsup_{n \to \infty} \frac{1}{\gamma} \ln E\left[e^{\gamma \eta^n + \gamma \int_{t \wedge \tau_n}^{\tau_n} \alpha_s d\langle M\rangle_s} \bigg| \mathcal{F}_t\right] \leq \frac{1}{\gamma} \ln E\left[e^{\gamma \eta + \gamma \int_t^T \alpha_s d\langle M\rangle_s} \bigg| \mathcal{F}_t\right]. \tag{43}$$

On the other hand applying the *sup* property and Lemma 3.5 we get that

$$\limsup_{n \to \infty} \operatorname*{ess\,sup}_{u \in U} E\left[\mathcal{E}_{t \wedge \tau_n, \tau_n}\left(\int f'_l(u) dM\right)\left(\eta^n + \int_{t \wedge \tau_n}^{\tau_n} g(s, u_s) d\langle M\rangle_s\right) \bigg| \mathcal{F}_t\right] \geq$$

$$\operatorname*{ess\,sup}_{u \in U} \limsup_{n \to \infty} E\left[\mathcal{E}_{t \wedge \tau_n, \tau_n}\left(\int f'_l(u) dM\right)\left(\eta^n + \int_{t \wedge \tau_n}^{\tau_n} g(s, u_s) d\langle M\rangle_s\right) \bigg| \mathcal{F}_t\right] =$$

$$= E\left[\mathcal{E}_{t,T}\left(\int f'_l(u) dM\right)\left(\eta + \int_t^T g(s, u_s) d\langle M\rangle_s\right) \bigg| \mathcal{F}_t\right] = V_t. \tag{44}$$

So from (42),(43) and (44) we obtain the upper bound for value process V:

$$V_t \leq \frac{1}{\gamma} \ln E\left[e^{\gamma \eta + \gamma \int_t^T \alpha_s d\langle M\rangle_s} \bigg| \mathcal{F}_t\right].$$

For the lower bound insert the control $u^0 \equiv 0$ in the representation of V:

$$E\left[\mathcal{E}_{t,T}\left(\int f'_l(0)dM\right)\left(\eta + \int_t^T f(s,0)d\langle M\rangle_s\right)\Big|\mathcal{F}_t\right] \leq V_t.$$

Finally we get an a priori estimate for the value process V:

$$E\left[\mathcal{E}_{t,T}\left(\int f'_l(0)dM\right)\left(\eta + \int_t^T f(s,0)d\langle M\rangle_s\right)\Big|\mathcal{F}_t\right] \leq V_t \leq \frac{1}{\gamma}\ln E\left[e^{\gamma\eta + \gamma\int_t^T \alpha_s d\langle M\rangle_s}\Big|\mathcal{F}_t\right].$$

Because $\eta + \int_0^T f(s,0)d\langle M\rangle_s \geq -C$, we have

$$V_t + \int_0^t f(s,0)d\langle M\rangle_s \geq E\left[\mathcal{E}_{t,T}\left(\int f'_l(0)dM\right)\left(\eta + \int_0^T f(s,0)d\langle M\rangle_s\right)\Big|\mathcal{F}_t\right] \geq -C$$

and from $|f(s,0)| \leq \alpha_s$ we obtain the following estimate for the value process:

$$-C - \int_0^t \alpha_s d\langle M\rangle_s \leq V_t \leq \frac{1}{\gamma}\ln E\left[e^{\gamma\eta + \gamma\int_t^T \alpha_s d\langle M\rangle_s}\Big|\mathcal{F}_t\right].$$

This implies the estimate for $V^+ + \int \alpha d\langle M\rangle$:

$$V_t^+ + \int_0^t \alpha_s d\langle M\rangle_s \leq \frac{1}{\gamma}\ln E\left[e^{\gamma\eta^+ + \gamma\int_t^T \alpha_s d\langle M\rangle_s}\Big|\mathcal{F}_t\right] + \int_0^t \alpha_s d\langle M\rangle_s =$$

$$= \frac{1}{\gamma}\ln E\left[e^{\gamma\eta^+ + \gamma\int_t^T \alpha_s d\langle M\rangle_s}\Big|\mathcal{F}_t\right] + \frac{1}{\gamma}\ln E\left[e^{\gamma\int_0^t \alpha_s d\langle M\rangle_s}\Big|\mathcal{F}_t\right] \leq$$

$$\leq \frac{1}{\gamma}\ln E\left[e^{\gamma\eta^+ + \gamma\int_0^T \alpha_s d\langle M\rangle_s}\Big|\mathcal{F}_t\right].$$

Using the latter estimate for $V^+ + \int \alpha d\langle M\rangle$ and Doob's martingale inequality we will have:

$$Ee^{p(V^+ + \int \alpha d\langle M\rangle)^*} \leq Ee^{\frac{p}{\gamma}\ln\left(E\left[e^{\gamma\eta^+ + \gamma\int_0^T \alpha_s d\langle M\rangle_s}\big|\mathcal{F}_t\right]\right)^*} =$$

$$= E\left|E\left[e^{\gamma\eta^+ + \gamma\int_0^T \alpha_s d\langle M\rangle_s}\big|\mathcal{F}_t\right]^*\right|^{\frac{p}{\gamma}} \leq$$

$$\leq C_p E\left(e^{\gamma\eta^+ + \gamma\int_0^T \alpha_s d\langle M\rangle_s}\right)^{\frac{p}{\gamma}} = C_p Ee^{p\eta^+ + p\int_0^T \alpha_s d\langle M\rangle_s} < \infty.$$

For the lower bound we have: $V_t^- \leq C + \int_0^t \alpha_s d\langle M\rangle_s$ and $(V^-)^* \leq C + \int_0^T \alpha_s d\langle M\rangle_s$. Now because $Ee^{p(V^-)^*} \leq Ee^{pC + p\int_0^T \alpha_s d\langle M\rangle_s} < \infty$ we obtain that $V \in \aleph$. □

We prove the following uniqueness theorem using the method developed by F. Delbaen, Y. Hu and A. Richou (2009) [11], where the uniqueness of equation (1) was proved in the case of Brownian filtration.

Theorem 3.3. If there exists a solution of equation (1) from the class $\aleph$ then it is unique.

Proof. Define the new class of controls:

$$\tilde{U} = \left\{ (u_t)_{0 \leq t \leq T} \; : \; E\mathcal{E}_T\left(\int f_l'(u)dM\right) = 1 \; ; \; E^u\left[|\eta| + \int_0^T |f(s,u_s) - f_l'(s,u_s)u_s|d\langle M\rangle_s\right] < \infty \right\}$$

and the corresponding value process:

$$\tilde{V}_t = \text{ess sup}_{u \in \tilde{U}} E^u\left[\eta + \int_t^T [f(s,u_s) - f_l'(s,u_s)u_s]d\langle M\rangle_s \Big| \mathcal{F}_t\right].$$

We shall prove that if Y is a solution of equation (1) from the class $\aleph$ then $Y_t = \tilde{V}_t$ for any $t \in [0;T]$ P a. s. Let first show that for any $u \in \tilde{U}$

$$Y_t \geq E^u\left[\eta + \int_t^T [f(s,u_s) - f_l'(s,u_s)u_s]d\langle M\rangle_s \Big| \mathcal{F}_t\right].$$

For simplicity let us make notation: $g(s,x) = f(s,x) - f_l'(s,x)x$.
Because Y is the solution of equation (1), applying the Ito formula and the inequality for convex functions it is easy to verify that

$$\mathcal{E}_t\left(\int f_l'(u)dM\right)\left(Y_t + \int_0^t g(s,u_s)d\langle M\rangle_s\right)$$

is a local supermartingale. Let $(\tau_n)_{n \geq 1}$ be the localization sequence for

$$\mathcal{E}_t\left(\int f_l'(u)dM\right)\left(Y_t + \int_0^t g(s,u_s)d\langle M\rangle_s\right).$$

Then by the supermartingale property we get for any $n \geq 1$

$$\mathcal{E}_{t \wedge \tau_n}\left(\int f_l'(u)dM\right)\left(Y_{t \wedge \tau_n} + \int_0^{t \wedge \tau_n} g(s,u_s)d\langle M\rangle_s\right) \geq$$

$$\geq E\left[\mathcal{E}_{\tau_n}\left(\int f_l'(u)dM\right)\left(Y_{\tau_n} + \int_0^{\tau_n} g(s,u_s)d\langle M\rangle_s\right) \Big| \mathcal{F}_t\right]$$

which is equivalent to the following inequality

$$Y_{t \wedge \tau_n} \geq E\left[\mathcal{E}_{t \wedge \tau_n, \tau_n}\left(\int f_l'(u)dM\right)\left(Y_{\tau_n} + \int_{t \wedge \tau_n}^{\tau_n} g(s,u_s)d\langle M\rangle_s\right) \Big| \mathcal{F}_t\right] =$$

$$= E^u\left[Y_{\tau_n} + \int_{t \wedge \tau_n}^{\tau_n} g(s,u_s)d\langle M\rangle_s \Big| \mathcal{F}_t\right]. \tag{45}$$

We need to take limits to the both sides of the inequality. For this we verify conditions for dominated convergence theorem. It is clear that

$$\left|\int_{t \wedge \tau_n}^{\tau_n} g(s,u_s)d\langle M\rangle_s\right| \leq \int_0^T |g(s,u_s)|d\langle M\rangle_s$$

and

$$E^u \int_0^T |g(s,u_s)|d\langle M\rangle_s < \infty$$

since $u \in \tilde{U}$.

For the family $(Y_{\tau_n})_{n \geq 1}$ it is sufficient to use the Fatou's Lemma. We have inequality $Y_{\tau_n} \geq -(Y^-)^*$. So we only need $E^u(Y^-)^*$ to be finite. To this end we shall prove the following lemma:

Lemma 3.12. $E\Big[\mathcal{E}_t\Big(\int f'_l(u)dM\Big)\ln\mathcal{E}_t\Big(\int f'_l(u)dM\Big)\Big] < \infty$ for any $t \in [0;T]$ and $u \in \tilde{U}$.

Proof. Define stopping times:

$$\sigma_n = inf\Big\{t \geq 0 : \int_0^t f'^2_l(s,u_s)d\langle M\rangle_s > n\Big\} \wedge T.$$

Using successively definition of $\mathcal{E}\Big(\int f'_l(u)dM\Big)$, Girsanov's Theorem, Lemma 3.17 (see Appendix) and the inequality

$$xy \leq e^{px} + \frac{y}{p}(\ln y - \ln p - 1), \qquad \forall (x,y) \in R \times R^+ \tag{46}$$

we obtain the following chain of inequalities:

$$E\Big[\mathcal{E}_{t\wedge\sigma_n}\Big(\int f'_l(u)dM\Big)\ln\mathcal{E}_{t\wedge\sigma_n}\Big(\int f'_l(u)dM\Big)\Big] = \frac{1}{2}E^u\Big[\int_0^{t\wedge\sigma_n} f'^2_l(s,u_s)d\langle M\rangle_s\Big] \leq$$

$$\leq \gamma E^u \int_0^{t\wedge\sigma_n} \alpha_s d\langle M\rangle_s + \gamma E^u \int_0^{t\wedge\sigma_n} |g(s,u_s)|d\langle M\rangle_s \leq$$

$$\leq \gamma E e^{p\int_0^{t\wedge\sigma_n} \alpha_s d\langle M\rangle_s} + \frac{\gamma}{p}E\Big[\mathcal{E}_{t\wedge\sigma_n}\Big(\int f'_l(u)dM\Big)\ln\mathcal{E}_{t\wedge\sigma_n}\Big(\int f'_l(u)dM\Big)\Big] -$$

$$-\frac{\gamma}{p}(1+\ln p) + \gamma E^u \int_0^{t\wedge\sigma_n} |g(s,u_s)|d\langle M\rangle_s.$$

This we can rewrite in the form:

$$\Big(1-\frac{\gamma}{p}\Big)E\Big[\mathcal{E}_{t\wedge\sigma_n}\Big(\int f'_l(u)dM\Big)\ln\mathcal{E}_{t\wedge\sigma_n}\Big(\int f'_l(u)dM\Big)\Big] \leq$$

$$\leq \gamma E e^{p\int_0^{t\wedge\sigma_n} \alpha_s d\langle M\rangle_s} + \gamma E^u \int_0^{t\wedge\sigma_n} |g(s,u_s)|d\langle M\rangle_s - \frac{\gamma}{p}(1+\ln p) \leq$$

$$\leq \gamma E e^{p\int_0^T \alpha_s d\langle M\rangle_s} + \gamma E^u \int_0^T |g(s,u_s)|d\langle M\rangle_s - \frac{\gamma}{p}(1+\ln p) < \infty. \tag{47}$$

On the other hand applying (47) and the Fatou's lemma we finish the proof

$$E\Big[\mathcal{E}_t\Big(\int f'_l(u)dM\Big)\ln\mathcal{E}_t\Big(\int f'_l(u)dM\Big)\Big] =$$

$$= E\lim_n\inf\mathcal{E}_{t\wedge\sigma_n}\Big(\int f'_l(u)dM\Big)\ln\mathcal{E}_{t\wedge\sigma_n}\Big(\int f'_l(u)dM\Big) \leq$$

$$\leq \lim_n\inf E\Big[\mathcal{E}_{t\wedge\sigma_n}\Big(\int f'_l(u)dM\Big)\ln\mathcal{E}_{t\wedge\sigma_n}\Big(\int f'_l(u)dM\Big)\Big] \leq$$

$$\leq \frac{p\gamma}{p-\gamma}E e^{p\int_0^T \alpha_s d\langle M\rangle_s} + \frac{p\gamma}{p-\gamma}E^u\int_0^T |g(s,u_s)|d\langle M\rangle_s - \frac{\gamma}{p-\gamma}(1+\ln p) < \infty.$$

□

Using inequality (46) and Lemma 3.12 we obtain

$$E^u(Y^-)^* \leq Ee^{\varepsilon(Y^-)^*} + \frac{1}{\varepsilon}E\left[\mathcal{E}_T\left(\int f_l'(u)dM\right)\ln \mathcal{E}_T\left(\int f_l'(u)dM\right)\right] - \frac{1+\ln \varepsilon}{\varepsilon} < \infty.$$

So using the Fatou's Lemma in (45) we obtain:

$$Y_t \geq E^u\left[\eta + \int_t^T g(s,u_s)d\langle M\rangle_s \Big| \mathcal{F}_t\right]$$

for any $u \in \tilde{U}$, hence $Y_t \geq \tilde{V}_t$.

Now we left to show the reverse inequality $Y_t \leq \tilde{V}_t$.

Let first prove

$$E\left[\mathcal{E}_t\left(\int f_l'(Z)dM\right)\ln \mathcal{E}_t\left(\int f_l'(Z)dM\right)\right] < \infty,$$

where Z is the second component of the solution of equation (1). Hence $\mathcal{E}_t(\int f_l'(Z)dM)$ is a uniformly integrable martingale. For this we define stopping times:

$$\tau_n = \inf\left\{t \geq 0 : \max\left(\int_0^t f_l'^2(s,Z_s)d\langle M\rangle_s; \int_0^t Z_s^2 d\langle M\rangle_s\right) > n\right\} \wedge T$$

and corresponding measures $dQ_n = \mathcal{E}_{\tau_n}(\int f_l'(Z)dM)dP$. Because (Y,Z,L) is a solution of equation (1) we have the equalities:

$$Y_{t\wedge\tau_n} = Y_0 - \int_0^{t\wedge\tau_n}[f(s,Z_s) - Z_s f_l'(s,Z_s)]d\langle M\rangle_s +$$

$$+ \int_0^{t\wedge\tau_n} Z_s(dM_s - f_l'(s,Z_s)d\langle M\rangle_s) + L_{t\wedge\tau_n}$$

and

$$E^{Q_n}\left[Y_{t\wedge\tau_n} + \int_0^{t\wedge\tau_n}[f(s,Z_s) - Z_s f_l'(s,Z_s)]d\langle M\rangle_s\right] = EY_0.$$

Using the same arguments as in Lemma 3.12 we obtain

$$E\left[\mathcal{E}_{t\wedge\tau_n}\left(\int f_l'(Z)dM\right)\ln \mathcal{E}_{t\wedge\tau_n}\left(\int f_l'(Z)dM\right)\right] = \frac{1}{2}E^{Q_n}\left[\int_0^{t\wedge\tau_n} f_l'^2(s,Z_s)d\langle M\rangle_s\right] \leq$$

$$\leq \gamma E^{Q_n}\left[\int_0^{t\wedge\tau_n} \alpha_s d\langle M\rangle_s\right] + \gamma E^{Q_n}\left[\int_0^{t\wedge\tau_n}[Z_s f_l'(s,Z_s) - f(s,Z_s)]d\langle M\rangle_s\right] =$$

$$= \gamma E^{Q_n}\left[\int_0^{t\wedge\tau_n} \alpha_s d\langle M\rangle_s\right] + \gamma E^{Q_n}[Y_{t\wedge\tau_n}] - \gamma EY_0 =$$

$$= \gamma E^{Q_n}\left[Y_{t\wedge\tau_n} + \int_0^{t\wedge\tau_n} \alpha_s d\langle M\rangle_s\right] - \gamma EY_0 \leq$$

$$\leq \gamma Ee^{p(Y_{t\wedge\tau_n} + \int_0^{t\wedge\tau_n} \alpha_s d\langle M\rangle_s)} - \frac{\gamma}{p}(1+\ln p) - \gamma EY_0 +$$

$$+ \frac{\gamma}{p}E\left[\mathcal{E}_{t\wedge\tau_n}\left(\int f_l'(Z)dM\right)\ln \mathcal{E}_{t\wedge\tau_n}\left(\int f_l'(Z)dM\right)\right].$$

This implies that

$$\left(1-\frac{\gamma}{p}\right)E\left[\mathcal{E}_{t\wedge\tau_n}\left(\int f'_l(Z)dM\right)\ln\mathcal{E}_{t\wedge\tau_n}\left(\int f'_l(Z)dM\right)\right] =$$

$$= \frac{1}{2}\left(1-\frac{\gamma}{p}\right)E^{Q_n}\left[\int_0^{t\wedge\tau_n} f_l'^{2}(s,Z_s)d\langle M\rangle_s\right] \le$$

$$\le \gamma E e^{p(Y_{t\wedge\tau_n} + \int_0^{t\wedge\tau_n} \alpha_s d\langle M\rangle_s)} - \frac{\gamma}{p}(1+\ln p) - \gamma EY_0 \le$$

$$\le \gamma E e^{p(Y^+ + \int \alpha d\langle M\rangle)^*} - \frac{\gamma}{p}(1+\ln p) - \gamma EY_0 < \infty. \qquad (48)$$

On the other hand applying Fatou's Lemma

$$E\left[\mathcal{E}_t\left(\int f'_l(Z)dM\right)\ln\mathcal{E}_t\left(\int f'_l(Z)dM\right)\right] =$$

$$= E\liminf_n \left[\mathcal{E}_{t\wedge\tau_n}\left(\int f'_l(Z)dM\right)\ln\mathcal{E}_{t\wedge\tau_n}\left(\int f'_l(Z)dM\right)\right] \le$$

$$\le \liminf_n E\left[\mathcal{E}_{t\wedge\tau_n}\left(\int f'_l(Z)dM\right)\ln\mathcal{E}_{t\wedge\tau_n}\left(\int f'_l(Z)dM\right)\right] \le$$

$$\le \frac{p\gamma}{p-\gamma} E e^{p(Y^+ + \int \alpha d\langle M\rangle)^*} - \frac{\gamma}{p-\gamma}(1+\ln p) - \frac{p\gamma}{p-\gamma} EY_0 < \infty.$$

Thus $\left\{\mathcal{E}_t \int f'_l(Z)dM\right\}_{t\in[0;T]}$ is a martingale and we can define the new probability measure $dQ = \mathcal{E}_T\left(\int f'_l(Z)dM\right)dP$.

Now we show that the process $Z \in \tilde{U}$. According to Fatou's lemma and inequality (48)

$$E^Q \int_0^T f_l'^{2}(s,Z_s)d\langle M\rangle_s = E^Q \liminf_n \int_0^{\tau_n} f_l'^{2}(s,Z_s)d\langle M\rangle_s \le$$

$$\le \liminf_n E^Q \int_0^{\tau_n} f_l'^{2}(s,Z_s)d\langle M\rangle_s < \infty.$$

This means that $\int_0^t f'_l(s,Z_s)dM_s - \int_0^t f_l'^{2}(s,Z_s)d\langle M\rangle_s$ is a square integrable Q martingale. Let us show that $E^Q\left[\int_0^T |f(s,Z_s) - f'_l(s,Z_s)Z_s|d\langle M\rangle_s\right] < \infty$. Using inequalities $f(s,Z_s) - f'_l(s,Z_s)Z_s \le f(s,0) \le \alpha_s$ (which follows from the convexity of f and from the quadratic growth condition) we get

$$E^Q\left[\int_0^T [f(s,Z_s) - f'_l(s,Z_s)Z_s]^+ d\langle M\rangle_s\right] \le E^Q\left[\int_0^T \alpha_s d\langle M\rangle_s\right] \le$$

$$\le E e^{p\int_0^T \alpha_s d\langle M\rangle_s} + \frac{1}{p}E\left[\mathcal{E}_T\left(\int f'_l(Z)dM\right)\ln\mathcal{E}_T\left(\int f'_l(Z)dM\right)\right] - \frac{1+\ln p}{p} < \infty. \qquad (49)$$

Notice that

$$E^Q\left[\int_0^T [f(s,Z_s) - f'_l(s,Z_s)Z_s]d\langle M\rangle_s\right] = E^Q\left[\int_0^T [f(s,Z_s) - f'_l(s,Z_s)Z_s]^+ d\langle M\rangle_s\right] -$$

$$-E^Q\left[\int_0^T [f(s,Z_s) - f'_l(s,Z_s)Z_s]^- d\langle M\rangle_s\right] = E^Q[Y_0 - Y_T]. \tag{50}$$

From (49) and (50)

$$E^Q\left[\int_0^T [f(s,Z_s) - f'_l(s,Z_s)Z_s]^- d\langle M\rangle_s\right] = E^Q[Y_T] - EY_0 +$$

$$+E^Q\left[\int_0^T [f(s,Z_s) - f'_l(s,Z_s)Z_s]^+ d\langle M\rangle_s\right] \leq E^Q\left[Y_T^+ + \int_0^T \alpha_s d\langle M\rangle_s\right] - EY_0 \leq$$

$$\leq E^Q\left[\left(Y^+ + \int \alpha d\langle M\rangle\right)^*\right] - EY_0 \leq$$

$$\leq Ee^{p\left(Y^+ + \int \alpha d\langle M\rangle\right)^*} - \frac{1 + \ln p}{p} - EY_0 +$$

$$+\frac{1}{p}E\left[\mathcal{E}_T\left(\int f'_l(Z)dM\right)\ln \mathcal{E}_T\left(\int f'_l(Z)dM\right)\right] < \infty.$$

This means that $E^Q\left[\int_0^T |f(s,Z_s) - f'_l(s,Z_s)Z_s| d\langle M\rangle_s\right] < \infty$. Now let us show that $E^Q|\eta| < \infty$. As we know $\eta^+ \leq (Y^+)^*$ and $\eta^- \leq C + \int_0^T \alpha_s d\langle M\rangle_s$. Therefore $E^Q[\eta^+] \leq E^Q(Y^+)^* < \infty$ and $E^Q[\eta^-] \leq C + E^Q\left[\int_0^T \alpha_s d\langle M\rangle_s\right] < \infty$. This implies that the process Z is from the class $\tilde{U}$. Now because $\int_0^t f'_l(s,Z_s)dM_s - \int_0^t f'^2_l(s,Z_s)d\langle M\rangle_s$ is a square integrable Q martingale we obtain from the Girsanov Theorem that

$$Y_t = E^Q\left[\eta + \int_t^T [f(s,Z_s) - f'_l(s,Z_s)Z_s]d\langle M\rangle_s \bigg| \mathcal{F}_t\right]$$

and $Y_t \leq \tilde{V}_t$, since $Z \in \tilde{U}$.

Remark. It is evident that $V_t = \tilde{V}_t$, since by Proposition 3.6 V is the solution of equation (1) from the class $\aleph$. □

It follows from the proof of the **Theorem 3.3** that if $Y \in \aleph$ then $\mathcal{E}(f'_l(Z) \cdot M)$ is a uniformly integrable martingale. It is interesting to ask whether is the exponential martingale $\mathcal{E}(Z \cdot M)$ uniformly integrable. The following proposition gives a partial answer on this question.

Proposition 3.7. *Let the generator f satisfies the quadratic growth condition $|f(s,x)| \leq \alpha_s + \frac{\gamma}{2}x^2$ and (Y,Z,L) is the solution of equation (1) such that*

$$Ee^{p(Y^+ + \int \alpha d\langle M\rangle)^*} < \infty$$

for some $p > \gamma$. Then for any $\beta \in (p - \sqrt{p(p-\gamma)}; p + \sqrt{p(p-\gamma)})$ $\mathcal{E}(\beta \int ZdM + \beta L)$ is a uniformly integrable martingale on $[0;T]$.

Proof. Define stopping times $\tau_n = \inf\left\{t \geq 0 : \int_0^t Z_s^2 d\langle M\rangle_s + \langle L\rangle_t \geq n\right\}$ and new probability measures $dQ_\beta^n = \mathcal{E}_{\tau_n}(\beta \int ZdM + \beta L)dP$ where $\beta > \frac{\gamma}{2}$. Our aim is to prove that the family $\{\mathcal{E}_{\tau_n}(\beta \int ZdM + \beta L)\}_{n\geq 1}$ is uniformly integrable, which implies that $\mathcal{E}(\beta \int ZdM + \beta L)$ is a uniformly integrable martingale. Because (Y,Z,L) is the solution of equation (1), we have

$$Y_{\tau_n} = Y_0 - \int_0^{\tau_n} f(s,Z_s)d\langle M\rangle_s + \beta \int_0^{\tau_n} Z_s^2 d\langle M\rangle_s + \beta \langle L\rangle_{\tau_n} +$$

$$+ \int_0^{T_n} Z_s(dM_s - \beta Z_s d\langle M\rangle_s) + L_{T_n} - \beta\langle L\rangle_{T_n}$$

and according to Girsanov's Theorem

$$E^{Q^n_\beta}[Y_{T_n} - Y_0] = E^{Q^n_\beta}\int_0^{T_n}[\beta Z_s^2 - f(s, Z_s)]d\langle M\rangle_s + \beta E^{Q^n_\beta}\langle L\rangle_{T_n} \geq$$

$$\geq E^{Q^n_\beta}\int_0^{T_n}[\beta Z_s^2 - \frac{\gamma}{2}Z_s^2 - \alpha_s]d\langle M\rangle_s + \beta E^{Q^n_\beta}\langle L\rangle_{T_n} =$$

$$= \left(\beta - \frac{\gamma}{2}\right)E^{Q^n_\beta}\int_0^{T_n} Z_s^2 d\langle M\rangle_s + \beta E^{Q^n_\beta}\langle L\rangle_{T_n} - E^{Q^n_\beta}\int_0^{T_n}\alpha_s d\langle M\rangle_s.$$

Since $\beta > \frac{\gamma}{2}$ we can rewrite the last inequality in the form

$$\frac{1}{2}E^{Q^n_\beta}\int_0^{T_n} Z_s^2 d\langle M\rangle_s + \frac{\beta}{2\beta-\gamma}E^{Q^n_\beta}\langle L\rangle_{T_n} \leq$$

$$\leq \frac{1}{2\beta-\gamma}E^{Q^n_\beta}\left[Y_{T_n} + \int_0^{T_n}\alpha_s d\langle M\rangle_s\right] - \frac{1}{2\beta-\gamma}EY_0$$

and because $\frac{\beta}{2\beta-\gamma} > \frac{1}{2}$ for $\beta > \frac{\gamma}{2}$

$$\frac{1}{2}E^{Q^n_\beta}\left[\int_0^{T_n} Z_s^2 d\langle M\rangle_s + \langle L\rangle_{T_n}\right] \leq$$

$$\leq \frac{1}{2\beta-\gamma}E^{Q^n_\beta}\left[Y_{T_n} + \int_0^{T_n}\alpha_s d\langle M\rangle_s\right] - \frac{1}{2\beta-\gamma}EY_0. \qquad (51)$$

Using successively definition of $\mathcal{E}(\beta \int ZdM + \beta L)$, Girsanov's theorem and inequalities (51) and (46) we obtain:

$$E\left[\mathcal{E}_{T_n}\left(\beta\int ZdM + \beta L\right)\ln\mathcal{E}_{T_n}\left(\beta\int ZdM + \beta L\right)\right] =$$

$$= \frac{\beta^2}{2}E^{Q^n_\beta}\left[\int_0^{T_n} Z_s^2 d\langle M\rangle_s + \langle L\rangle_{T_n}\right] \leq$$

$$\leq \frac{\beta^2}{2\beta-\gamma}E^{Q^n_\beta}\left[Y_{T_n} + \int_0^{T_n}\alpha_s d\langle M\rangle_s\right] - \frac{\beta^2}{2\beta-\gamma}EY_0 \leq$$

$$\leq \frac{\beta^2}{2\beta-\gamma}Ee^{p(Y_{T_n} + \int_0^{T_n}\alpha_s d\langle M\rangle_s)} - \frac{\beta^2}{2\beta-\gamma}EY_0 - \frac{1+\ln p}{p} +$$

$$+ \frac{\beta^2}{2\beta-\gamma}\cdot\frac{1}{p}E\left[\mathcal{E}_{T_n}\left(\beta\int ZdM + \beta L\right)\ln\mathcal{E}_{T_n}\left(\beta\int ZdM + \beta L\right)\right].$$

Let us rewrite last inequality in the form:

$$\left(1 - \frac{\beta^2}{2\beta-\gamma}\cdot\frac{1}{p}\right)E\left[\mathcal{E}_{T_n}\left(\beta\int ZdM + \beta L\right)\ln\mathcal{E}_{T_n}\left(\beta\int ZdM + \beta L\right)\right] \leq$$

$$\leq \frac{\beta^2}{2\beta-\gamma}Ee^{p(Y_{T_n} + \int_0^{T_n}\alpha_s d\langle M\rangle_s)} - \frac{\beta^2}{2\beta-\gamma}EY_0 - \frac{1+\ln p}{p}.$$

Now if $1 - \frac{\beta^2}{2\beta-\gamma} \cdot \frac{1}{p} > 0$ since

$$Ee^{p(Y_{\tau_n} + \int_0^{\tau_n} \alpha_s d\langle M \rangle_s)} \leq Ee^{p(Y^+ + \int \alpha d\langle M \rangle)^*} < \infty$$

we obtain that the family $\{\mathcal{E}_{\tau_n}(\beta \int Z dM + \beta L)\}_{n \geq 1}$ is uniformly integrable. So we need to choose $\beta > \frac{\gamma}{2}$ such that

$$\frac{\beta^2}{2\beta - \gamma} < p.$$

The last inequality is equivalent to the following $\beta^2 - 2p\beta + p\gamma < 0$ for which

$$\beta \in (p - \sqrt{p(p-\gamma)}; p + \sqrt{p(p-\gamma)}).$$

Finally because $p - \sqrt{p(p-\gamma)} > \frac{\gamma}{2}$ we obtain that the family $\{\mathcal{E}_{\tau_n}(\beta \int Z dM + \beta L)\}_{n \geq 1}$ is uniformly integrable for any $\beta \in (p - \sqrt{p(p-\gamma)}; p + \sqrt{p(p-\gamma)})$, hence $\mathcal{E}(\beta \int Z dM + \beta L)$ is a martingale. □

Remark. It is easy to verify that if $p > \gamma$ then $\gamma \in (p - \sqrt{p(p-\gamma)}; p + \sqrt{p(p-\gamma)})$, so if $\gamma = 1$ we obtain that $\mathcal{E}(\int Z dM + L)$ is uniformly integrable martingale.

3.7 Appendix: Some auxiliary assertions

Lemma 3.13. *If a convex generator f has quadratic growth with respect to an argument z, then the left-derivative f'_l of f will also have quadratic growth with respect to z. I.e. if there exist a predictable non-negative process α_t and a constant $\gamma \geq 0$ such that $|f(t, \omega, z)| \leq \alpha_t(\omega) + \frac{\gamma}{2} z^2$, then the following inequality holds too:*

$$|f'_l(t, \omega, z)| \leq \gamma + 2\alpha_t(\omega) + \frac{3\gamma}{2} z^2.$$

Proof. Because f is convex, for any $z_1 < z_2 < z_3$

$$\frac{f(t, \omega, z_2) - f(t, \omega, z_1)}{z_2 - z_1} \leq f'_l(t, \omega, z_2) \leq \frac{f(t, \omega, z_3) - f(t, \omega, z_2)}{z_3 - z_2}.$$

If we take $z_1 = z - 1$, $z_2 = z$, $z_3 = z + 1$, then we have the mutual inequality

$$f(t, \omega, z) - f(t, \omega, z - 1) \leq f'_l(t, \omega, z) \leq f(t, \omega, z + 1) - f(t, \omega, z).$$

So we can write the estimate of $|f'_l|$

$$|f'_l(t, \omega, z)| \leq \max\left(|f(t, \omega, z + 1) - f(t, \omega, z)|; |f(t, \omega, z) - f(t, \omega, z - 1)|\right).$$

Now let us estimate these two values:

$$|f(t, \omega, z) - f(t, \omega, z - 1)| \leq |f(t, \omega, z)| + |f(t, \omega, z - 1)|$$
$$\leq \alpha_t + \frac{\gamma}{2} z^2 + \alpha_t + \frac{\gamma}{2}(z - 1)^2$$
$$\leq 2\alpha_t + \frac{\gamma}{2} z^2 + \gamma(z^2 + 1) = \gamma + 2\alpha_t + \frac{3\gamma}{2} z^2.$$

In the same way we obtain that $|f(t, \omega, z + 1) - f(t, \omega, z)| \leq \gamma + 2\alpha_t + \frac{3\gamma}{2} z^2$. □

Let us recall that U is the class of predictable bounded controls.

Lemma 3.14. *If $|f(s,x)| \leq \alpha_s + \frac{\gamma}{2}x^2$ and M, $\int \alpha dM \in BMO$, then for any $u \in U$ $\int f'_l(u) dM \in BMO$.*

Proof. According to Lemma 3.13 $|f'_l(t,\omega,z)| \leq \gamma + 2\alpha_t(\omega) + \frac{3\gamma}{2}z^2$ and since $u \in U$, there exists a constant D such that $|u_s| \leq D$. So we have

$$\int_t^T |f'_l(s,u_s)|^2 d\langle M \rangle_s \leq \int_t^T \left|2\alpha_s + \gamma + \frac{3\gamma}{2}|u_s|^2\right|^2 d\langle M \rangle_s$$

$$\leq 3\int_t^T 4\alpha_s^2 d\langle M \rangle_s + 3\int_t^T \gamma^2 d\langle M \rangle_s + 3\int_t^T \frac{9\gamma^2}{4}|u_s|^4 d\langle M \rangle_s$$

$$\leq 12\int_t^T \alpha_s^2 d\langle M \rangle_s + 3\gamma^2(\langle M \rangle_T - \langle M \rangle_t) + \frac{27\gamma^2}{4}D^4(\langle M \rangle_T - \langle M \rangle_t).$$

Because $\int \alpha dM \in BMO$ and $M \in BMO$ we obtain that $\int f'_l(u)dM \in BMO$. $\square$

Lemma 3.15. *Let U_n be the class of predictable bounded with n controls: $U_n = \{u \in U : |u_s(\omega)| \leq n\}$. Then for any a.s. finite and predictable ψ process the following equality holds:*

$$\operatorname*{ess\,sup}_{u \in U_n}[f(s,u_s) + f'_l(s,u_s)(\psi_s - u_s)] = f(s,\psi^n_s) + f'_l(s,\psi^n_s)(\psi_s - \psi^n_s) = g^n(s,\psi_s),$$

where $\psi^n_s = n\mathbf{1}_{(\psi_s > n)} + \psi_s \mathbf{1}_{(|\psi_s| \leq n)} - n\mathbf{1}_{(\psi_s < -n)}$ and

$$g^n(s,x) = \mathbf{1}_{(x > n)}[f(s,n) + f'_l(s,n)(x - n)] + \mathbf{1}_{(|x| \leq n)}f(s,x) + \mathbf{1}_{(x < -n)}[f(s,-n) + f'_l(s,-n)(x + n)].$$

Proof. Let us consider three cases:

1) If $|\psi_s| \leq n$, then we have

$$f(s,\psi^n_s) + f'_l(s,\psi^n_s)(\psi_s - \psi^n_s) = f(s,\psi_s) \geq f(s,u_s) + f'_l(s,u_s)(\psi_s - u_s) \quad \forall u \in U_n.$$

2) If $\psi_s > n$, then because f is convex for any $u \in U_n$ the following inequality holds:

$$f(s,u_s) - f(s,n) + f'_l(s,u_s)(\psi_s - u_s)$$

$$\leq f(s,u_s) - f(s,n) + \frac{f(s,n) - f(s,u_s)}{n - u_s} \cdot (\psi_s - u_s)$$

$$= (f(s,n) - f(s,u_s))\left(\frac{\psi_s - u_s}{n - u_s} - 1\right)$$

$$= \frac{f(s,n) - f(s,u_s)}{n - u_s} \cdot (\psi_s - n) \leq f'_l(s,n)(\psi_s - n)$$

so when $\psi_s > n$, we have for any $u \in U_n$:

$$f(s,u_s) + f'_l(s,u_s)(\psi_s - u_s) \leq f(s,n) + f'_l(s,n)(\psi_s - n).$$

3) If $\psi_s < -n$, then again because f is convex for any $u \in U_n$ we have

$$f(s,u_s) - f(s,-n) + f'_l(s,u_s)(\psi_s - u_s)$$

$$\leq f(s,u_s) - f(s,-n) + \frac{f(s,u_s) - f(s,-n)}{n + u_s} \cdot (\psi_s - u_s)$$

$$= (f(s,u_s) - f(s,-n))\left(1 + \frac{\psi_s - u_s}{n + u_s}\right)$$

$$= \frac{f(s,u_s) - f(s,-n)}{n + u_s} \cdot (\psi_s + n) \leq f'_l(s,-n)(\psi_s + n).$$

So, when $\psi_s < -n$, for any $u \in U_n$ the following inequality holds:
$$f(s, u_s) + f'_l(s, u_s)(\psi_s - u_s) \leq f(s, -n) + f'_l(s, -n)(\psi_s + n).$$

Based on these three cases, we can conclude that the following inequality is fulfilled:
$$f(s, \psi_s^n) + f'_l(s, \psi_s^n)(\psi_s - \psi_s^n) \geq f(s, u_s) + f'_l(s, u_s)(\psi_s - u_s) \quad \forall u \in U_n.$$

On the other hand, because $\psi^n \in U_n$, it is obvious that
$$f(s, \psi_s^n) + f'_l(s, \psi_s^n)(\psi_s - \psi_s^n) \leq \operatorname{ess\,sup}_{u \in U_n}[f(s, u_s) + f'_l(s, u_s)(\psi_s - u_s)].$$

This means that we have the equality
$$\operatorname{ess\,sup}_{u \in U_n}[f(s, u_s) + f'_l(s, u_s)(\psi_s - u_s)] = f(s, \psi_s^n) + f'_l(s, \psi_s^n)(\psi_s - \psi_s^n). \qquad \square$$

Lemma 3.16. *For any a.s. finite and predictable process φ the following equality holds:*
$$\operatorname{ess\,sup}_{u \in U}[f(s, u_s) + f'_l(s, u_s)(\varphi_s - u_s)] = f(s, \varphi_s).$$

Proof. According to Lemma 3.15, we have the equalities
$$\operatorname{ess\,sup}_{u \in U}[f(s, u_s) + f'_l(s, u_s)(\varphi_s - u_s)]$$
$$= \sup_n \left\{ \operatorname{ess\,sup}_{u \in U_n}[f(s, u_s) + f'_l(s, u_s)(\varphi_s - u_s)] \right\}$$
$$= \lim_n \operatorname{ess\,sup}_{u \in U_n}[f(s, u_s) + f'_l(s, u_s)(\varphi_s - u_s)]$$
$$= \lim_n g^n(s, \varphi_s) = f(s, \varphi_s). \qquad \square$$

Lemma 3.17. *If convex generator f satisfies quadratic growth condition $|f(s, x)| \leq \alpha_s + \frac{\gamma}{2}x^2$, then the following estimate for a squire of the left derivative of f holds true:*
$$\frac{1}{2}f'^2_l(s, y) \leq \gamma\alpha_s + \gamma(yf'_l(s, y) - f(s, y)).$$

Proof. From the convexity of f for any x and y we have the inequality: $f(s, y) + f'_l(s, y)(x - y) \leq f(s, x)$. From this and the quadratic growth condition of f we easily deduce that
$$xf'_l(s, y) \leq f(s, x) + yf'_l(s, y) - f(s, y) \leq \alpha_s + \frac{\gamma}{2}x^2 + yf'_l(s, y) - f(s, y).$$

Now if we insert $x = \frac{1}{\gamma}f'_l(s, y)$ in the last inequality, we obtain
$$\frac{1}{\gamma}f'^2_l(s, y) \leq \alpha_s + (yf'_l(s, y) - f(s, y)) + \frac{1}{2\gamma}f'^2_l(s, y)$$

and from this it is elementary to deduce the inequality
$$\frac{1}{2}f'^2_l(s, y) \leq \gamma\alpha_s + \gamma(yf'_l(s, y) - f(s, y)).$$

$\square$

4 Application to the linear-quadratic regulator problem and relation with Bellman-Chitashvili equation

Consider the linear-quadratic regulator problem (LQR):
Let A be a decision set and let M be continuous local martingale. To any $a \in A$ we associate a local martingale $M^a = aM$. Controls are predictable mappings $u : \Omega \times [0;T] \longrightarrow A$ and probability measures P^u corresponding to any control u are defined by $dP^u = \mathcal{E}_T(M^u)dP$, where $M_t^u = \int_0^t u_s dM_s$, provided that $\mathcal{E}_t(M^u)$ is a uniformly integrable martingale. Assume that the cost criterium is of the form
$$r(t,a) = -g(t)a^2 + h(t)$$
and consider an optimization problem to maximize
$$E\Big[\mathcal{E}_T\Big(\int udM\Big)\Big(\eta + \int_0^T [h(s) - g(s)u_s^2]d\langle M\rangle_s\Big)\Big] \qquad (52)$$
over all $u \in U$, where η is a $\mathcal{F}_T$-measurable random variable and the class of controls U is following:
$$U = \Big\{u_t : E\mathcal{E}_T\Big(\int udM\Big) = 1\,;\, E^u\Big[|\eta| + \int_0^T |h(s) - g(s)u_s^2|d\langle M\rangle_s\Big] < \infty\Big\}.$$
Let
$$V_t = \operatorname{ess\,sup}_{u \in U} E^u\Big[\eta + \int_t^T [h(s) - g(s)u_s^2]d\langle M\rangle_s \Big| \mathcal{F}_t\Big]$$
be the value process of the problem.

We can formally derive the corresponding BSDE for the value process using R. Chitashvili's [9] result, who derived equation for V for a general system of martingales $(M^a; a \in A)$.

Theorem (Chitashvili [9]) Let M^a be a family of continuous local martingales such that $\langle M^a \rangle_t \ll K_t$, $t \in [0;T]$ for some predictable, increasing process K. Assume that the Radon-Nikodim derivative $\frac{d\langle M^a \rangle_t}{dK_t}$ is continuous w.r.t. a $dK_t \times dP$ a. s. and let
$$\int_0^t \max_{a \in A} \frac{d\langle M^a \rangle_s}{dK_s} dK_s \in A_{loc}^+. \qquad (53)$$
Then the value process
$$V_t = \operatorname{ess\,sup}_{u \in U} E^u\Big[\eta + \int_t^T r(s,u_s)dK_s \Big| \mathcal{F}_t\Big], \quad t \in [0;T]$$
is the solution of the equation
$$\begin{cases} V_t = V_0 - \int_0^t \max_{a \in A}\Big[r(s,a) + \frac{d\langle m, M^a\rangle_s}{dK_s}\Big]dK_s + m_t, \\ V_T = \eta. \end{cases} \qquad (54)$$

The solution of (54) is the couple (V,m) where V is a semimartingale, m is a local martingale and (V,m) satisfies equation (54).

In our case $A = R$, $r(s,a) = -g(s)a^2 + h(s)$, $M^a = aM$ and $K = \langle M \rangle$. Applying the Galtchouk-Kunita-Watanabe decomposition for the local martingale m we obtain:

$$m_t = \int_0^t Z_s dM_s + L_t \tag{55}$$

where Z is a predictable, M integrable process and L is a local martingale orthogonal to M. Using successively all of this we get:

$$\max_{a \in A} \left[r(s,a) + \frac{d\langle m, M^a \rangle_s}{dK_s} \right] = \max_{a \in R} \left[-g(s)a^2 + h(s) + \frac{d\langle \int Z dM + L, aM \rangle_s}{d\langle M \rangle_s} \right] =$$

$$= \max_{a \in R} \left[-g(s)a^2 + h(s) + aZ_s \right] = h(s) + \frac{1}{4g(s)} Z_s^2.$$

So in our case equation (54) has the following form:

$$\begin{cases} Y_t = Y_0 - \int_0^t \left[h(s) + \frac{1}{4g(s)} Z_s^2 \right] d\langle M \rangle_s + \int_0^t Z_s dM_s + L_t, \\ Y_T = \eta. \end{cases} \tag{56}$$

Note that in this case condition (53) is not satisfied and the existence of a solution of corresponding BSDE does not follows from the R. Chitashvili's theorem ([9]). But we can use the theorems of existence and uniqueness of the solution which was proved in the previous section. Note that in case of equation (56) generator has the form: $f(s,z) = h(s) + \frac{z^2}{4g(s)}$ and satisfies quadratic growth condition

$$|f(s,z)| \le |h(s)| + \frac{1}{4\varepsilon} z^2 = |h(s)| + \frac{\gamma}{2} z^2$$

where for all s $g(s) \ge \varepsilon > 0$ and $\gamma = \frac{1}{2\varepsilon}$. So according to the **Theorems 3.1 and 3.3** under the conditions:

(i) $M, \int |h| dM \in BMO$

(ii) $Ee^{p(\eta^+ + \int_0^T |h(s)| d\langle M \rangle_s)^*} < \infty$ and $\eta + \int_0^T h(s) d\langle M \rangle_s \ge -D > -\infty$

there exists the unique solution of equation (56) $(\tilde{V}, \varphi, L)$ where the first component is represented in the form:

$$\tilde{V}_t = \operatorname{ess\,sup}_{u \in \tilde{U}} E \left[\mathcal{E}_{t,T} \left(\int \frac{u}{2g} dM \right) \left(\eta + \int_t^T \left[h(s) - \frac{u_s^2}{4g(s)} \right] d\langle M \rangle_s \right) \middle| \mathcal{F}_t \right] \tag{57}$$

and the class of controls $\tilde{U}$ is following:

$$\tilde{U} = \left\{ u_t : E\mathcal{E}_T \left(\int \frac{u}{2g} dM \right) = 1 ; E \left[\mathcal{E}_T \left(\int \frac{u}{2g} \right) \left(|\eta| + \int_0^T \left| h(s) - \frac{u_s^2}{4g(s)} \right| d\langle M \rangle_s \right) \right] < \infty \right\}.$$

It was also proved in **Theorem 3.3** that the second component of the triple $(\tilde{V}, \varphi, L)$ belongs to the class $\tilde{U}$ and is optimal. This means that $\tilde{V}$ can be represented as follows:

$$\tilde{V}_t = E \left[\mathcal{E}_{t,T} \left(\int \frac{\varphi}{2g} dM \right) \left(\eta + \int_t^T \left[h(s) - \frac{\varphi_s^2}{4g(s)} \right] d\langle M \rangle_s \right) \middle| \mathcal{F}_t \right]. \tag{58}$$

Theorem 4.1. If the conditions (i) and (ii) are satisfied, then the predictable process $\frac{\varphi}{2g}$ is the optimal control for the optimization problem (52).

Proof. It is clear that $u \in \tilde{U}$ implies $\frac{u}{2g} \in U$ and $u \in U$ implies $2gu \in \tilde{U}$. This means that the process $\frac{\varphi}{2g}$ belongs to the class U. So we only need to show that $\frac{\varphi}{2g}$ maximizes (52). Let $u \in U$. This implies that $2gu \in \tilde{U}$. So according to (57) and (58) we obtain:

$$E\left[\mathcal{E}_T\left(\int \frac{\varphi}{2g}dM\right)\left(\eta + \int_0^T (h(s) - \frac{\varphi_s^2}{4g(s)})d\langle M\rangle_s\right)\right] \geq$$

$$\geq E\left[\mathcal{E}_T\left(\int \frac{2gu}{2g}dM\right)\left(\eta + \int_0^T (h(s) - \frac{(2g(s)u_s)^2}{4g(s)})d\langle M\rangle_s\right)\right] =$$

$$= E\left[\mathcal{E}_T\left(\int udM\right)\left(\eta + \int_0^T (h(s) - g(s)u_s^2)d\langle M\rangle_s\right)\right]$$

for any $u \in U$. This means that $\frac{\varphi}{2g}$ is the optimal solution for the optimization problem (52). □

For the generality we can consider the case when $M^a = M^0 + aM$, where M^0 is a local martingale orthogonal to M. Because M and M^0 are orthogonal local martingales we obtain that

$$\mathcal{E}_t(M^u) = \mathcal{E}_t\left(M^0 + \int udM\right) = \mathcal{E}_t(M^0) \cdot \mathcal{E}_t\left(\int udM\right).$$

So if $\mathcal{E}(M^0)$ is a martingale, then the optimization problem (52) can be written with respect to the measure $d\tilde{P} = \mathcal{E}_T(M^0)dP$:

$$E\left[\mathcal{E}_T\left(M^0 + \int udM\right)\left(\eta + \int_0^T [h(s) - g(s)u_s^2]d\langle M\rangle_s\right)\right] =$$

$$= \tilde{E}\left[\mathcal{E}_T\left(\int udM\right)\left(\eta + \int_0^T [h(s) - g(s)u_s^2]d\langle M\rangle_s\right)\right] \quad (59)$$

over all $u \in U$, where the class of controls U is following:
$u \in U$, if and only if, $\mathcal{E}(\int udM)$ is an uniformly integrable $\tilde{P}$-martingale and

$$\tilde{E}\left[\mathcal{E}_T\left(\int udM\right)\left(|\eta| + \int_0^T |h(s) - g(s)u_s^2|d\langle M\rangle_s\right)\right] < \infty.$$

Note that $\langle M^a\rangle_t \ll \langle M\rangle_t + \langle M^0\rangle_t \equiv K_t$ $t \in [0;T]$ for any $a \in R$. To use Chitashvili's theorem ([9]) we need to rewrite the optimization problem (59) with respect to the process K:

$$\sup_{u \in U} \tilde{E}\left[\mathcal{E}_T\left(\int udM\right)\left(\eta + \int_0^T [h(s) - g(s)u_s^2]d\langle M\rangle_s\right)\right] =$$

$$= \sup_{u \in U} \tilde{E}\left[\mathcal{E}_T\left(\int udM\right)\left(\eta + \int_0^T [h(s) - g(s)u_s^2]\frac{d\langle M\rangle_s}{dK_s}dK_s\right)\right]$$

and the value process will have the form:

$$V_t = \operatorname*{ess\,sup}_{u \in U} \tilde{E}\left[\mathcal{E}_{t,T}\left(\int udM\right)\left(\eta + \int_t^T [h(s) - g(s)u_s^2]\frac{d\langle M\rangle_s}{dK_s}dK_s\right)\Big|\mathcal{F}_t\right].$$

Now using the decomposition (55) in this case we obtain that

$$\max_{a \in R}\left[r(s,a)\frac{d\langle M\rangle_s}{dK_s} + \frac{d\langle m, M^a\rangle_s}{dK_s}\right] =$$

$$= h(s)\frac{d\langle M\rangle_s}{dK_s} + \frac{1}{4g(s)}Z_s^2\frac{d\langle M\rangle_s}{dK_s} + \frac{d\langle L, M^0\rangle_s}{dK_s}$$

and the equation (54) will have the following form:

$$\begin{cases} Y_t = Y_0 - \int_0^t \left[h(s) + \frac{1}{4g(s)}Z_s^2\right]d\langle M\rangle_s + \int_0^t Z_s dM_s + L_t - \langle L, M^0\rangle_t, \\ Y_T = \eta. \end{cases} \quad (60)$$

According to the Girsanov theorem M and $L - \langle L, M^0\rangle$ are orthogonal $\tilde{P}$ local martingales, so we only need to solve equation (60) with respect to the measure $\tilde{P}$. So if the following conditions

(ĩ) $M, \int |h|dM \in BMO(\tilde{P})$

(ĩĩ) $\tilde{E}e^{p(\eta^+ + \int_0^T |h(s)|d\langle M\rangle_s)^*} < \infty$ and $\eta + \int_0^T h(s)d\langle M\rangle_s \geq -D > -\infty$

are satisfied, then according to the **Theorem 3.1** and **Theorem 3.3** there exists the unique solution $(\tilde{V}, \tilde{\varphi}, \tilde{L})$ of equation (60) and the following theorem is true:

Theorem 4.1' Let M^0 be a continuous local martingale orthogonal to M, such that $\mathcal{E}(M^0)$ is a martingale. Then if (ĩ) and (ĩĩ) holds true the predictable process $\frac{\tilde{\varphi}}{2g}$ is the optimal control for the optimization problem (59).

References

[1] S. Ankirchner, P. Imkeller and G. Reis, "Classical and variational differentiability of bsdes with quadratic growth", *Electronic Journal of Probability* **12** (2007), 1418–1453.

[2] P. Barrieu, N. Cazanave and N. El Karoui, "Closedness results for BMO semi-martingales and application to quadratic BSDEs," *Comptes Rendus Mathematique* **346** (2008), 881–886.

[3] J.-M. Bismut, Conjugate convex functions in optimal stochastic control, *J. Math. Anal. Appl.* **44** (1973), 384–404.

[4] J. M. Bismut, "Controle des systemes lineaires quadratiques: applications de l'integrale stochastique," *Seminaire de Probabilites XII (eds.: C. Dellacherie, P. A. Meyer, and M. Weil), Lecture Notes in Mathematics 649*, Springer-Verlag, Berlin/Heidelberg (1978), 180–264.

[5] P. Briand and Y. Hu, Quadratic BSDEs with convex generators and unbounded terminal conditions, *Probab. Theory Related Fields* **141**:3-4 (2008), 543–567.

[6] B. Chikvinidze, "Backward stochastic differential equations with a convex generator," *Georgian Mathematical Journal* **19** (2012), 63–92.

[7] B. Chikvinidze, Semimartingale Backward Equations with Convex Generator, *BULLETIN OF THE GEORGIAN NATIONAL ACADEMY OF SCIENCES*, (2011), vol. 5, no. 3.

[8] B. Chikvinidze and M. Mania "ON THE GIRSANOV TRANSFORMATION OF BMO MARTINGALES," *Reports of Enlarged Session of the Seminar of I. Vekua Institute of Applied Mathematics*, (2012), Volume 26.

[9] R. Chitashvili, Martingale ideology in the theory of controlled stochastic processes, in: *Probability theory and mathematical statistics (Tbilisi, 1982)*, 73–92, Lecture Notes in Math., 1021, Springer, Berlin, 1983.

[10] R. Chitashvili and M. Mania, Optimal locally absolutely continuous change of measure. Finite set of decisions. I, *Stochastics* **21**:2 (1987), 131–185.

[11] F. Delbaen, Y. Hu, and A. Richou, On the uniqueness of solutions to quadratic BSDEs with convex generators and unbounded terminal conditions, *Ann. Inst. Henri Poincare Probab. Stat.* **47**:2 (2011), 559–574.

[12] F. Delbaen, P. Monat, W. Schachermayer, M. Schweizer and C. Stricker, "Weighted norm inequalities and hedging in incomplete markets," *Finance and Stochastics* **1** (1997) 181–227.

[13] F. Delbaen and S. Tang, "Harmonic analysis of stochastic equations and backward stochastic differential equations," *Probability Theory and Related Fields* **146** (2010) 291–336.

[14] C. Dellacherie and P.-A. Meyer, *Probabilities and Potential. B. Theory of Martingales*, Translated from the French by J. P. Wilson. North-Holland Mathematics Studies, 72. North-Holland Publishing Co., Amsterdam, 1982.

[15] C. Doleans-Dade and P. A. Meyer, "Inegalites de normes avec poids," *Universite de Strasbourg Seminaire de Probabilites*, XIII, (1979) 313–331.

[16] N. El Karoui, Les aspects probabilistes du controle stochastique, in: *Ninth Saint Flour Probability Summer School – 1979 (Saint Flour, 1979)*, pp. 73–238, Lecture Notes in Math., 876, Springer, Berlin-New York, 1981.

[17] R. J. Elliott, *Stochastic Calculus and Applications*, Applications of Mathematics (New York), 18. Springer-Verlag, New York, 1982.

[18] C. Frei, M Mocha and N. Westray, "BSDEs in Utility Maximization with BMO Market Price of Risk", *Working Paper*, arXiv:1107.0183v1, 2011.

[19] Y. Hu, P. Imkeller, and M. Müller, "Utility maximization in incomplete markets," *Annals of Applied Probability* **15** (2005) 1691–1712.

[20] N. Kazamaki, "On transforming the class of BMO-martingales by a change of law," *Tohoku Mathematical Journal* **31** (1979) 117–125.

[21] N. Kazamaki, *Continuous Exponential Martingales and BMO*, Lecture Notes in Mathematics, 1579. Springer-Verlag, Berlin, 1994.

[22] M. Kobylanski, Backward stochastic differential equations and partial differential equations with quadratic growth, *Ann. Probab.* **28**:2 (2000) 558–602.

[23] M. Kohlmann and S. Tang, "Minimization of risk and linear quadratic optimal control theory," *SIAM Journal on Control and Optimization* **42** (2003) 1118–1142.

[24] H. Kunita and S. Watanabe, On square integrable martingales, *Nagoya Math. J.* **30** (1967), 209–245.

[25] J. P. Lepeltier and J. San Martin, Existence for BSDE with superlinear-quadratic coefficient, *Stochastics Stochastics Rep.* **63**:3-4 (1998) 227–240.

[26] M. Mania and M. Schweizer, "Dynamic exponential indifference valuation," *Annals of Applied Probability* **15** (2005) 2113–2143.

[27] M. Mania and R. Tevzadze, "A semimartingale Bellman equation and the variance-optimal martingale measure," *Georgian Mathematical Journal* **7** (2000) 765–792.

[28] M. Mania and R. Tevzadze, "Martingale equation of exponential type," *Electronic communication in probability* **11** (2006) 206–216.

[29] M. Mania, M. Santacroce and R. Tevzadze, "A semimartingale BSDE related to the minimal entropy martingale measure", *Finance and Stochastics* **7** No. 3, (2003) 385–402.

[30] Mocha, M., Westray, N.: Quadratic semimartingale BSDEs under an exponential moments condition. 382 In: Seminaire de Probabilites XLIV, Lecture Notes in Mathematics pp. 105139 (2012)

[31] M. A. Morlais, Quadratic BSDEs driven by a continuous martingale and applications to the utility maximization problem, *Finance Stoch.* **13**:1 (2009), 121–150.

[32] E. Pardoux and S. G. Peng, Adapted solution of a backward stochastic differential equation, *Systems Control Lett.* **14**:1 (1990), 55–61.

[33] R. Tevzadze, Solvability of backward stochastic differential equations with quadratic growth, *Stochastic Process. Appl.* **118**:3 (2008), 503–515.

[34] W. Schachermayer, "A characterization of the closure of H_∞ in BMO," *Seminaire de Probabilites XXX, Lecture Notes in Mathematics 1626*, Springer, Berlin, (1996) 344-356.

i want morebooks!

Buy your books fast and straightforward online - at one of world's fastest growing online book stores! Environmentally sound due to Print-on-Demand technologies.

Buy your books online at
www.get-morebooks.com

Kaufen Sie Ihre Bücher schnell und unkompliziert online – auf einer der am schnellsten wachsenden Buchhandelsplattformen weltweit! Dank Print-On-Demand umwelt- und ressourcenschonend produziert.

Bücher schneller online kaufen
www.morebooks.de

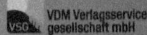 VDM Verlagsservicegesellschaft mbH
Heinrich-Böcking-Str. 6-8   Telefon: +49 681 3720 174   info@vdm-vsg.de
D - 66121 Saarbrücken   Telefax: +49 681 3720 1749 www.vdm-vsg.de

www.ingramcontent.com/pod-product-compliance
Lightning Source LLC
Chambersburg PA
CBHW031958240526
45464CB00024B/1182